TIME BOMB

HOW TO SURVIVE
THE UPCOMING
ICELANDIC VOLCANIC
ERUPTIONS

Mark Reed

iUniverse, Inc.
Bloomington

TIME BOMB
HOW TO SURVIVE THE UPCOMING
ICELANDIC VOLCANIC ERUPTIONS

iUniverse books may be ordered through booksellers or by contacting:

iUniverse
1663 Liberty Drive
Bloomington, IN 47403
www.iuniverse.com
1-800-Authors (1-800-288-4677)

ISBN: 978-1-4620-0979-4 (sc)
ISBN: 978-1-4620-0980-0 (ebk)

Printed in the United States of America

iUniverse rev. date: 3/28/2011

Contents

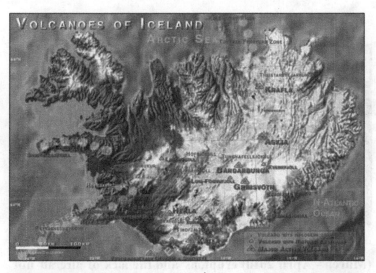

"Map of Iceland showing all major volcanoes and fissure rift areas."

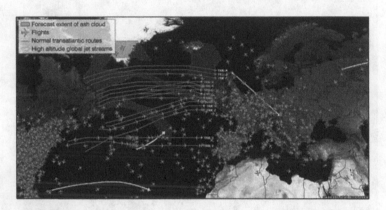

"A map showing the ash fall out zone from the recent (March - April 2010) eruptons and the lack of aircraft not flying in the no fly zones which followed the eruptions."

"This book is dedicated to my loving wife Lisa, my beautiful daughter Valentine and my wonderful son Harry for all the support, help and inspiration you have all given me over the years"

How to survive the upcoming Icelandic eruptions

Introduction

By Mark Reed

I have always been interested in natural disasters, like hurricanes, tornadoes, tsunamis and other natural phenomena. I enjoy watching the Discovery Channel with my son Harry as a relaxing time after work, sometimes.

When all the media frenzy occurred in March 2010 regarding the Icelandic volcanic eruption it was as if the Discovery Channel was now on every news channel, every time I picked up a newspaper or turned on the TV, there it was Eyjafjallajokull the longest hardest sounding word I have ever heard of and the bane of many a newsreader who must have had nightmares each night at having to endlessly announce the continuing eruption of the impossibly sounding volcano from Iceland.

I was amazed that a seemingly smallish eruption 1000 miles from the UK and Europe could cause such disruption to everyone's lives, even my son Harry was telling me that one of his teachers from school was trapped in the USA and it took her five weeks to get back to work. One of my closest friend's son was trapped in Perugia in Italy and after incessant phone calls and discussions, his son had to get trains, buses, and minicabs costing my friend a pretty penny or two to get him home to Blighty. As all this mayhem continued and I saw Michael O' Leary's (Ryanair's MD) unhappy face moaning about compensating thousands of stranded travellers for an act of god that wasn't his fault.

I started thinking about what would happen to all of us in Europe should a much larger volcano blow its top, how the effects would be magnified, how people would cope with being separated from their families and friends, how would we get food from the shops, water to drink, electricity, gas and all the other thousands of issues that make up our complicated lifestyles in the 21st century. I then was browsing through one of the various volcano websites on the net, as one does after work sometimes and I came across Laki, a huge strato volcano and fissure volcano over in Iceland, the fateful eruptions of 1783 and 1784 when most of Europe and the UK was covered by a yellowish poison gas cloud from an Icelandic volcano and approximately 6 million people died worldwide and it possibly also caused the French Revolution. The more I read I was flabbergasted, I didn't know anything about this eruption which occurred about 230 years ago and I have had a good education and have been reading encyclopaedia's and other reference books since I was 12 and read newspapers or books every single day on my way to and from work on the underground between Hendon Central and Waterloo on the blessed Northern Line, since I have had my latest job down in Clapham Junction. I am going through a decent sized novel a week and am probably keeping Waterstones afloat single handed.

How could something so important be hidden away in some technical papers from some Icelandic Institute of Volcanology and no one know about this eruption that didn't occur in Java or Indonesia but here in the UK and Europe. The more I read the more determined I was that I would write a book about all the Icelandic volcanoes and there massive effect on Icelandic and European and world history over the last 10,000 years. It seemed that many major events in world history were shaped by volcanic eruptions and yet no one seemed to know much about it. We all know that Vesuvius blew up in 79 ACE and covered Pompeii with 20 metres of volcanic ash and the sad curled up figures of those burnt or roasted alive citizens of Pompeii in the ash are there for all the tourists to gawp at, however that was 2,000 years ago, not today. Some of us know that Santorini, or Thera as it was known, blew up in 1600 BCE and destroyed the Minoan Civilization, some of you may even have heard about Krakatoa which blew up with a mighty bang in 1883 in Java, however that is probably the extent of our knowledge of this subject.

Some of the Icelandic volcanoes have caused massive devastation across Europe and have affected weather patterns and possibly caused mini ice ages across Europe for thousands of years, quite a few of them have had such huge eruptions that volcanic ash has fallen from Norway to Helsinki and Ireland. Frankly I am amazed to find anyone still living in Iceland, they are surely hardy people descended from extremely tough and resourceful Vikings, no wonder we Anglo Saxons were terrified of the Vikings when they descended on the UK in the 8th Century ACE.

So I was driven to write my book, always checking the news to see if another volcano had blown its top. I felt that if I helped people prepare for the really big eruption that was bound to come sooner or later that at least I could say later, well I did what I could, it's up to people after reading my book to at least make an educated decision as to whether or not they want to make some provision for the definite next huge Icelandic eruption, which as sure as night follows day will occur one day, month or year very soon. I have used various source material for this book including Wikipedia the on line encyclopaedia which I think is a great resource tool for anyone writing a book, magazine articles or newspaper report, I have also used much information regarding how people in Britain coped during the Second World War, although we won't be bombed by the Luftwaffe or our ships sunk by the Kriegsmarine this time, the effects could be similar so I have mentioned occasionally how our grandparents coped 70 years ago with shortages of food, clothes, fuel and other hardships that they faced.

I would like to take this opportunity to thank my long suffering wife Lisa for putting up with me for the last 7 months it has taken to write this book and for bringing me endless supplies of lemon tea to keep me going. My son Harry has contributed the makers of some of the children's medicine's and has encouraged me to write the book as then I am near his bedroom so he can hear me tapping away on the computer each night whilst he falls asleep. I would also like to thank my daughter Valentine who I sent a sampler of the book to and she was amazed that I could actually string a few sentences together and when she actually read it properly she was very encouraging and gave me much moral support. Lastly and not least I would like to thank my mother Annette who lives in Limoges, France (still within striking distance of Laki's poisonous gases unfortunately), who commented that it was a little depressing and it needed to be more cheerful and positive, cheers mum, I did try and put a happy ending on the book though !!

Very finally and definitely lastly I would like to thank all my wife's six sisters and two brothers and mother in law who have teased me at every opportunity about buying cans of baked beans and rude associated jokes, this has indeed spurred me on to ensure that I make a success of the book, just so I can invite them to the European launch of the book and see them all go green when it becomes a best seller, and yes I will personally deliver a copy of my book to each of them with a can of beans attached.

My fervent hope is that this book will not be needed in the next few years, however my research tells me that volcanoes in Iceland do erupt with monotonous regularity about every three years, so it may be dear reader that you end up reading this book by torch or gas light from your darkened living room, whilst tucking into a can of cold baked beans, and you can bet your bottom dollar that I will be sitting in my house eating my cold beans saying to my wife, those six famous words that no wife likes to hear, "I told you to buy more beans darling." However when Katla, Laki or Bardarbunga goes up I guarantee there will be no beans, or any other foods on the shelves of Tesco, ASDA or Sainsbury's to buy so you had better stock up whilst you still can, read on for more information on "how to survive the upcoming Icelandic eruption". Have a look at my website www.icelandicvolcanoes.co.uk with regular Icelandic volcanic updates posted on it.

Mark Reed 2nd November 2010

ACKNOWLEGEMENTS

I would like to thank Ms Stacia A Langden, Archives Collection Manager of the Smithsonian Institute's Global Volcanism Program, for helping me with the acknowledgements for the photos of the various volcanoes shown in this book.

Individual acknowledgements must go to, Lee Siebert, The Nordic Volcanological Center, Michael Ryan at USGS and Oddur Sigurdson of the Iceland National Energy Authority

Section 1

How to survive the upcoming Icelandic volcanic eruptions

By Mark Reed

CHAPTER 1

What's going to hit us first?

Recently, no one living in Europe or North America could have not noticed the disruption caused by the ash cloud from the Icelandic Volcano with the unpronounceable name called Eyjafjallajokull, (let's call our friend EJ for the purposes of this book and saving gallons of black ink as well). I have done some research regarding this and other Icelandic Volcanoes and other volcanoes around the world together with the increasing incidence of earthquakes, hurricanes, typhoons and other phenomenon and I have come to the conclusion that the Earth is possibly going through some sort of cyclical movement which may have been foreseen by ancient cultures such as The Mayan, Egyptian and Chinese Civilizations and other ancient cultures.

If you throw evidence of global warming, the CO2 spike and speeding up of the melting of glaciers in Europe, Greenland and Asia, together with the breaking off of great ice sheets in Antarctica which have stood firm for 50,000 years or more then one can only come to the conclusion that things on Earth are about to change, probably not for the better.

Recently on the news I have heard that Eyjafjallajokull (EJ's) big brother Katla is possibly coming to life, it has

11

been said that the ash cloud and eruptions from Katla could be about 10 times more powerful and disruptive than Eyjafjallajokull. We will discuss these implications in a later chapter though.

My plan is that people throughout the world should slowly be starting to prepare for the worst eventuality so as to be prepared for the day when the newscaster comes on CNN and says with a solemn voice, "Breaking News, we have just heard that Mount Laki in Iceland has erupted and we have been told that the size of the eruption is 50 times more powerful than Eyjafjallajokull which caused so much disruption 2 years ago.....". I am not advocating that everyone rush down to the supermarket and buy all the canned food and bottled water and clearing out all the shops. No one really knows when the next super volcano on mega earthquake or giant tsunami will hit, however if I was a betting man I would say that within the next 5 years something pretty bad is bound to happen to our planet. I am not just talking about natural disasters either, we have had the largest upheaval in the worlds stock markets since 1929 and I think phase two, will be a crisis of confidence regarding sovereign debt, such as Ireland, Greece, Portugal and Spain will possibly cause the Euro to collapse and dismantle the EU, this will also cause major social upheavals throughout the world which could have a similar effect as a natural disaster as people's savings turn to nothing, pensions collapse and personal suffering increases to the point where civil strife and revolution against the existing regime's takes to the streets of Europe.

Another looming problem could easily be oil prices, a secret Pentagon report for the State Department has concluded that by 2015 that oil products e.g. petrol and heating oil will rise due to increased demand from especially India and China to a point that would mean that the ordinary man in the street would not be able to purchase petrol as the price would be ridiculously high e.g. £ 12.00 per gallon or similar.

Where does all this leave "The man on the Clapham Omnibus" the proverbial "man in the street" i.e. you and me, this leaves us up the creek without a paddle, or as the poets would kindly say up to our necks in it. If a major Icelandic eruption occurs we won't be getting Government handouts of food, water, medicines, and blankets initially because there would probably be nothing really that the Government would or could do for us. I am speaking here of the UK Government as I am a UK citizen, however what I am writing in this short book could well be transposed to any other country in the World. Only the most foolhardy of people would actually believe that governments would have enough (or any) provisions stashed away to feed, heat, water and keep going a population of 62,000,000 souls (the approximate current population of the UK). We are going to have to fend for ourselves and this book is hopefully going to be something of an emergency survival guide to keep you and your family safe and sound for between three months and nine months, which in all probability will be the length of time required to get matters back on track again.

Firstly, let's dwell for a while on what sort of catastrophe might be heading our way and try to plan to prepare for the most likely eventuality, It would be a great mistake planning for a two year freeze up, if we then had a tsunami which requires different planning criteria.

Meteor or Asteroid Strike

NASA has teams of astronomers with banks of computers constantly scanning the night skies for signs of·a comet, asteroid or meteor that could be heading our way. Everyone knows about the asteroid the size of Manhattan which supposedly wiped out the dinosaurs 65 million years ago. However personally I doubt that within the next 50 years or so anything life threatening is on course for a direct strike against planet earth. I predict that in 50 years time humans would definitely have the technology to strike back and either tow, incinerate or deflect any massive stellar object that comes too close to earth.

There has been a report that an asteroid will come fairly close to earth in 2029, however when the figures were recalculated it appears that this asteroid will miss the earth by a decent margin. However as long as you are not underneath the asteroid when it strikes and as long as it not as large as The Isle of Wight my advice in this book should help you to survive a medium strike on another continent or in the sea somewhere else on our planet.

On my 1% to 100 % scale (where 100 % is a definite hit) I would put Asteroid strike down at 20 % chance of a hit, fairly low.

I do not consider a major asteroid / meteor / comet strike as a real threat as if there was one coming the astronomers would have been yelling about it from the roof tops for some time now and as stated before my planning regime should cover this scenario.

Plague / Swine Flu /Sars / Biological Attack

The recent swine flu non event gave us some idea of how hard it must be to properly foresee what sort of flu pandemic or epidemic might be out there mutating right now in some pigsty in Bolivia or Peru.

If one looks back on serious pandemics in history, especially the "Spanish Flu Pandemic" of 1918 – 1919, there were strange combinations of factors which made the pandemic spread like wildfire. Some say it started in the unhealthy mire's of the trenches in World War 1 when many men were herded together in unhygienic conditions and as they were demobbed and sent home the epidemic went home with them. To be honest apart from locking oneself in one's home for 3 months and riding out the storm there is not much anyone can do about not getting infected in a pandemic.

My plans however for preparing for the worst would actually mean that a family could actually close their front door, batten down the hatches and survive for a couple of months thereby letting the pandemic burn itself out without getting contagion from contact with other persons.

A pandemic is survivable if we realise that it is coming and don't happen to be at the epicentre where the pandemic is first detected.

As a risk factor from 1 to 100 percent with 100 being 100 % sure it's going to happen, I put pandemic down at 28 % again a fairly low probability as I personally do not think we are all going to be wiped out by a pandemic in the next few years.

Crust Displacement Theory (see the Film 2012)

This is actually a theory of Albert Einstein's originally and is basically a theory that the Earth's crust is like an orange peel around an orange, if the peel is detached from the orange then it could slip around the orange whilst the orange was staying still. In essence the continental plates whizz around the earth, the magnetic poles reverse and all hell breaks loose on Earth (I bet they didn't really have those huge ships (Arks) built by the Chinese up in this Tibetan Mountains from the Film 2012).

They have found mammoths fully frozen in Siberia with mouthfuls of plants and flowers that they maintain grew thousands of miles away in warmer climes thus proving that it did happen in the last ice age over 10,000 years ago.

However current scientific thought is that if this does happen it will take thousands of years to move and we will not notice this move. On a scale of 1 – 100, I put this one down at a mere 5 % probability.

Tsunami's (Mega Tidal Waves)

Tsunami's can be caused by different mechanisms, the violent Boxing Day Tsunami of 2004 was caused by the

Indonesian Tectonic Plate lifting 30 meters in about 10 seconds and causing a huge sheet of elevated water to cascade over Banda Aceh, Indonesia, Thailand, India, Sri Lanka and as far away as Somalia and Ethiopia. To be honest if you live near a fault on the ocean floor i.e. Washington State in Western USA or on a coastline that is prone to tsunami activity, i.e. Japan, and the Pacific Rim Countries, USA (Western Seaboard), Hawaii, and some other Pacific Islands, then apart from moving or ensuring that the local authorities have built tsunami proof sea walls or proper 15 minute alarm warning systems then there is really not much you can do apart from moving to higher ground.

There is another form of tsunami which is, to my mind more dangerous than the first type, which is because there would be little or no warning at all if the second type of tsunami strikes you, this is the "landslide tsunami." These are caused by massive landslide's pouring into the sea or lakes which then can set off massively high tsunami's, sometimes nearly 400 metre's high, as in the Alaskan Tsunami of 1959 when a large landslide in a fjord type lake triggered a hugely high tsunami out of nowhere and a small fishing boat found itself hurtling along about 300 metres above the tree tops, in that case miraculously the fishing boat landed safely and the mariner told his strange tale of what happened.

There is an Island in the Canary Islands off the coast of North Africa called La Palma, the volcano is called Cumbre Vieja, which is an active volcano, there seems to be a major crack running across the Island and the theory goes that if water which is percolating down inside the volcano gets super heated and blown through fissure's in this volcano, this water action could lubricate the fissures to such an extent that half of the island would slip into the North Atlantic creating a tsunami that would then rush across the Atlantic Ocean at about 500 miles an hour and then three hours later a one kilometre high tsunami would then devastate the East Coast of USA up to thirty miles inland. The effects of this "mega tsunami" would be earth changing to say the least. In the UK although we might be protected from the worst effects of this disaster there would still be severe coastal flooding, the Thames Barrier would probably be breached and parts of Central London around Docklands, Westminster and Chelsea and Battersea would be under about ten foot of water, thousands of people would be homeless, all transport services including Underground, buses, taxis, main line stations would be at a standstill and possibly power supplies would be disrupted for some considerable time.

The volcano in The Canary Islands is not the only risk to the UK, geologists have discovered that massive landslides in prehistoric times in the Palaeolithic Era in Norway caused massive "mega tsunami" strikes against the Scottish Highlands and coasts of Northern England and Southern Scotland, possibly going inland for some miles.

Obviously if one was in the direct line of the tsunami, then there is not much that can be done by anyone, however if you lived ten miles or more inland or on high ground near the coast my ideas would still benefit most of the survivors of the tsunami as most public services could grind to a halt following such a strike.

Another well known scenario is that of a storm in the North Atlantic or North Sea combined with a high spring tide causing the sea to rise as in 1953 when the Isle of Thanet and much of the East Coast of England was under water, many lives were lost. The Dutch Dykes failed and much of Holland was flooded as well. Even with The Thames Barrier in place, London is sinking (as is most of Southern England) by about a half centimetre a year and the seas are rising by about the same amount per annum, eventually the UK will be forced to build sea walls and dykes on a huge scale to stop the majority of Southern England slipping under the waves, although this might be 100 to 200 years in the future, still a perfect storm brewing in the North Sea could well devastate the East Coast of England and seriously flood London if it overtops the Thames Barrier. The Thames Barrier used to be raised about twice a year about 10 years ago, now it is being raised about 10 times a year to protect London.

The chances of this scenario playing out on my 1 – 100 % scale would be about 50 % over the next 10 years, so the probability stakes are now rising considerably compared to meteors, pandemics and crust displacement theories.

Hurricane's, Typhoons and Severe Weather Conditions over the UK

The rhyme used in the film "My Fair Lady" goes, "In Hertford, Hereford and Hampshire, hurricanes hardly ever happen", which is mostly true as hurricane's and typhoons (as they are known in the Far East) are a tropical phenomenon. Hurricane's are formed when tropical thunder storms over West Africa drift over the

Atlantic in late summer, normally August onwards and due to the Earths centrifugal force (called the Correolis Effect) winds in The Tropics tend to blow anticlockwise in the Northern Hemisphere and clockwise in the Southern Hemisphere. Therefore these tropical storms blow into the Atlantic off Senegal and Togo in the hump off West Africa and then pick up moisture from very warm seas and therefore grow at the same time the Earths centrifugal force starts to turn the storm around upon itself, slowly a spinning motion is set up, the storm develops into "super cells" of heated, moisture laden clouds and these can in time turn into huge hurricane's which then bear down on the hapless islands of the Caribbean and Central America. Recently it was noted in 2006 that one hurricane followed another across the Atlantic and eventually about five hurricane's and two tropical storms had been tracked across the Caribbean in one hurricane season, an unheard of situation, possibly due to slightly warmer seas caused by global warming. In 2005 one memorable hurricane hit New Orleans and devastated the city (Hurricane Katarina)which is only starting to get back to its former self about now.

Global Warming, especially warmer seas are going to cause havoc with the world's weather systems over the next 10 – 50 years, many changes will be taking place that we here in the UK will see in our lifetimes. The warmer seas might allow the odd hurricane to blow along the North Atlantic and the remnants might even reach the UK. However more likely are local storms like the devastating storm which devastated much of Southern England in 1987. Many parts of the UK were affected by that storm with wind speeds up to nearly 100 mph or more in gusts. That was a once in 100 years storm, with global warming these storms might be an annual event. Local weather might cause problems too, there was a report of a small tornado in Cricklewood in London a couple of years ago and another in Birmingham as well.

However all these events will pale into insignificance if the "Gulf Stream" is turned off, this is the flow from the "Atlantic Conveyor" system which is part of a system of ocean currents which circles the globe and dissipates heat from The Tropics and warms up northern seas (Northern Atlantic Ocean) allowing much of Western Europe to remain fairly warmish even in the depth of winter. London is on the same latitude as New York, Moscow and Vancouver, however London very rarely sees its temperature sink below − 2 deg centigrade in winter, however Moscow can sink as low as − 40 deg centigrade and New York can be − 9 degrees. The reason for this is that the "Gulf Stream" transports millions of tons of warm water from the Caribbean up to Northern Norway and therefore right past the UK and Western Europe keeping us all fairly warm, it is the equivalent of thousands of power stations pumping out warm water into the North Atlantic, however there is now a big problem looming.

The Greenland Ice Cap and parts of the Canadian and Russian Tundra (Arctic Areas) are melting fairly quickly at present (whether you believe in global warming or not this is an irrefutable fact), this means that much more fresh water is pouring into the North Atlantic, the consequences of this are that this huge amount of fresh water is reducing the amount of salinity in the North Atlantic which stops the water from sinking down and running back southwards to complete the loop of the Atlantic Conveyer system which powers the Gulf Stream. If the Gulf Stream slows or stops (see the Film, "The Day After Tomorrow") then the UK and Europe will be in the grip of a "Mini Ice Age" much like Late Medieval Europe when the River Thames froze over on a regular basis and frost fairs were held on the Thames at least until the early Victorian Era. There is also the risk as well as the UK becoming more like Russia and Canada and experiencing continental extremely cold and snowy winters with the temperature in Southern England below − 10 for weeks, if not months on end and furthermore that Europe could briefly be tipped into a proper "Little Ice Age II", however this also depends on a lot of other factors including giant ash clouds from Icelandic Volcanoes cooling the Northern Hemisphere further and thereby magnifying the cooling effect from the disappearing Gulf Stream. There are many variant scenarios which could occur here including huge ice storms coming in from the North Sea and frozen sea areas off the Coast of England and Scotland. What this will do to Britain's ageing and creaking power grid, transformers, capacitors, rail network, water pipes, some of which are still Victorian and Edwardian is anybody's guess, we could be encountering weeks of sub zero temperature's with regular power cuts, interruptions to gas and oil supplies and much misery for the UK's population.

The 1 − 100 % chance of some variant of this scenario playing out within the next 10 years are about 58 %, if I were a gambling man I would be buying my thermal underwear from M & S tomorrow.

Icelandic Volcanic Activity Affecting the UK, Europe, USA and Canada

I want to tell a short story here about what happened in Europe in 1783 – 1784.. In Iceland in that year(1783) a volcanic fissure in the Earth erupted, this fissure or volcanic fissure was called Laki and together with the neighbouring volcanoes of Grimsvotn and Thordarhyrna ejected 14 Km3 or (3.4 cubic miles) of basalt lavas and clouds of hydrofluoric acid and sulphur dioxide, the gases were partly caused due the eruptions taking place through glacial waters and streams which caused further chemical reactions and huge amounts of gases and steam to be released into the plume of the eruptions.

Initially about 50 % of the livestock in Iceland was killed by the gases (fogs) and mostly due to starvation about 25 % of Iceland's population died due to the eruptions, and this was just the beginning of this disaster.

Unfortunately for Europe and the UK the prevailing winds from Iceland that year were south easterly and therefore these toxic fogs then started to blow across Europe. The gas clouds rose about 15 kilometre's high and it is estimated that eight million tons of hydrogen fluoride and 120 million tons of sulphur dioxide were eventually ejected, now called "The Laki Haze".

Firstly in the late spring of 1783 the haze hit Norway, Prague, Berlin, and then most of Europe including France reaching Britain on 22nd June 1783. In Britain it was known as the "Sand Summer". From local reports at the time the summer of 1783 was the hottest on record, there were reports from all over the UK of thick fogs and haze and that the sun was blood red, farm workers were dropping dead in the fields and mortality rates were up between 5 – 10 %, from newspaper reports of that year, 1783, from Newcastle Upon Tyne, it was reported that there were alternating days of massive thunderstorms with huge amounts of deafening lightning strikes and deaths of many horses and other animals from lightning strikes, then torrential rain and hail, then the fogs would descend again and outdoor workers were still dropping dead in the fields as people choked on the sulphur dioxide, which made their soft internal tissue swell and then caused death.

It is estimated that twenty three thousand people died in the UK in the period from June – December 1783. The hot summer fogs were so bad that many ships stayed in port not being able to even navigate outside of their home ports. This was probably apart from The English Civil War of the 1640's and the 1715 & 1745 Jacobite Rebellions and the Blitz of 1940 – 1941 the worst disaster to befall Britain in the last 230 years and yet no one seems to know about this true story.

The story does not end there, in France there were disastrous crop failures in 1785 – 1788 which left many peasants and farm workers starving, these people then went into the cities for food handouts and especially to Paris and this is definitely one of the causes of the French Revolution, as Marie Antoinette, King Louis XVI wife imperiously said "if there is no bread let them eat cake", however I doubt if there was any cake available let alone bread for many of these starving peasants who were then whipped up into an anti monarchy frenzy by Robespierre, Danton and Murat and other leaders of the French Revolution.

The story continues, the atmospheric movements of air due to these volcanic eruptions possibly affected the onset of the monsoons and rains in Africa and the Far East. It is reported that in Egypt the Nile fell to very low levels and as much as one sixth of the population died of starvation. In India as many as two million people died due to the non appearance of the Autumn Monsoon Rains and as far away as Japan many thousands perished in the Tenmei Famine of 1783 – 1784. This is all the result of one of Iceland's Volcanic Systems, there are 18 volcanic systems in Iceland, it sits over a hotspot in the Earth's Crust and strategically sits over the top end of the Mid Atlantic Rift which is a spreading rift zone between two continental plates.

Laki is just one of many bubbling volcanoes in Iceland that could blow at any moment and no one seems interested or cares what sort of damage, suffering and devastation these volcanoes could cause to the UK, Europe and around the world.

I have been browsing various volcano related websites and have been reading about Katla which is a massive volcano linked to Eyjafjallajokull, it is said that when EJ erupts normally Katla is not far behind and erupts as well. Katla (according to the Iceland Tourist Websites) has a 30 km crater hidden under a glacier, ideal for making another killer fog, Katla has been erupting intermittently as far back as the 14th century ACE, one of the most recent and huge eruptions took place in 1918 when the heat detached huge icebergs from the glacier which then floated out to sea. As recently as 1999 underground collapses of the glacier and localised flooding due to geothermal activity was taking place, this to my mind is one of 18 powder kegs ready to explode and we in the UK are sitting right in the middle of the cross hairs of the next "Laki Haze Eruption."

Not wanting to belabour the dangerousness of volcanic eruptions and associated disasters around the world, however at this juncture in this missive I would like to briefly touch upon "super volcanoes" which are different to the Icelandic and Hawaiian Volcanoes which sit on "hot spots" on tectonic fault lines. Tambora in Indonesia is one of a few fearsome "super volcanoes" which today occupies a vast lake in Indonesia, however 75,000 years ago when humans were just attempting to move out of Africa to populate the world, Tambora blew apart in a cataclysmic eruption that caused sulphuric acid rain, tsunami's, tephra ejections and it is estimated nearly wiped out human kind, scientists have estimated that there were probably only about 5000 humans left after Tambora erupted. Therefore hidden within the human psyche I would pose the thought that a fear of volcanic eruptions is somehow hard wired into the human sub conscious. Just look at the number of TV programmes, films and documentaries about volcanoes that have been distributed over the past few years, it is one of the film producers regular genre's after horror movies and crime movies.

Tambora has erupted again in 1815 – 1816, "the year without a summer" in Europe when famine stalked the continent again as the crops failed (as in 1783 – 8). This was the summer when Lord Byron invited his friend the poet and writer Shelly and Shelly's wife Mary Shelley to an old gothic chateau on Lake Constance and it is said that the terrific thunderstorms and lighting shows caused by Tambora's global ash cloud inspired Mary Shelley to write the novel "Frankenstein".

One of the other major "super volcanoes" is Yellowstone where the caldera (now sunken) is about the size of Greater London, recently the tilt meters on site have registered movement as the huge magma chamber underneath Yellowstone, (approximately 20 cubic kilometre's of magma) starts building up a head of steam, so to speak, it has been calculated that Yellowstone blows its top every 640,000 years, the last eruption (that re arranged lines of hills around the caldera) took place 640,000 years ago, so again we are all living on borrowed time. The earth would be plunged into a 2 to 3 year "winter" with hardly any sunlight if Yellowstone blows. The "sphereoids" (little pieces of molten red hot rock that gets ejected into the upper atmosphere by the eruption and then fall back to earth as small burning pieces of molten rock) would rain down over 1000 miles from the eruption, very soon all air traffic globally would shut down, no food would be able to be grown for several years, this is really one of the worst case scenarios that we will have to plan for.

Apart from "super volcanoes" there are also huge basalt lava flows that in earths past have spewed thousands of square miles of basalt lava over parts of the earth, possibly causing mass extinctions throughout the globe, such areas as India's "Deccan Traps" and Russia's "Siberian Traps" come to mind. Although volcanologists do not think we are currently in danger of one of these style lava flows.

In my 1% - 100 % probability charts for some sort of volcanic problem to hit the UK and Europe within the next 5 – 10 years I would state that I would put this one down to an 70 % probability, which in scientific terms means, start heading for the hills when you have finished eating your dinner. Hopefully my little book will at least give the average man in the street (and woman and children in the street) a fighting chance to survive this coming storm.

CHAPTER 2

How Volcanoes Have
Affected Human History

Throughout recorded history and even in prehistoric times it appears that volcanoes have affected the development of human civilizations to a much greater extent than we previously had thought. The sceptics amongst my readers may feel that with our 21st century advanced technology, gadgets and controls on the environment that we are the masters of our world, this is exactly how previous civilizations thought they were their own masters, we were wrong then and unfortunately I feel that we will be proved wrong again the next time a super volcano blasts off. I will in this chapter give some further examples of how volcanic eruptions have changed the course of history throughout recorded time.

A. Hekla (Iceland) Major eruptions approx 5000 BCE

Using tree ring dendranology and Irish Peat Bog tree ring dating techniques, archaeologists have deduced that a major eruption in Iceland affected weather patterns across Europe for some years possibly causing a drop in temperature's of between 5 – 8 degrees. This would equate to the UK going through its own "Little Ice Age" and would have severely affected the flora and fauna and human habitation at the time of the eruption. Most of Western Europe would have been affected by this eruption and it would have been one of the largest eruptions in the Holocene Era(the post ice age era that we are living in now).

B. Santorini (Thera) Mega Eruption approx 1600 BCE

This volcano which blew itself apart in the Bronze Age potentially caused history to be made and to be ended on many levels.

Firstly it is thought that the earthquakes, tephra ejections and mega tsunami's created by this massive volcano situated between Greece and Crete in the Eastern Mediterranean Sea caused the destruction of the Minoan (Cretan) Empire of the Middle Bronze Age. Palaces and buildings have been found along the Cretan Coast which was buried up to the ceiling with tephra and pumice stone and ash. The Minoan Civilization was technically and artistically well in advance of most of its neighbours and who knows what it could have grown into, another Roman Empire perhaps if it hadn't been destroyed by the huge Theran explosive volcano of 1600 BCE.

Plato in his histories mentions an island with concentric rings around it and a vast empire flourishing around this time, namely Atlantis which Plato states was destroyed in a day by cataclysmic eruptions and earthquakes until it slid beneath the waves. Who knows if a similar civilization to the

Minoan civilization was based on Santorini and was unfortunately based at the epicentre of the explosive Theran Eruption?

We also hear in The Bible that in Egypt Moses was asking Pharaoh to, "let my people go" and when Pharaoh would not agree to free his Hebrew Slaves, Moses then predicted that Egypt would be smitten by 10 plagues, it has been mooted by scientists over the last 30 years that many of the plagues sound like manifestations of volcanic influenced phenomena, for instance;

i. The river Nile turned to blood

It is a fact that due to seismic activity large amounts of CO_2 (carbon dioxide) can build up in lakes and rivers and when the lake or river is saturated with CO_2 then the colour turns red. Egypt is on a fault line very near the East African Rift Fault and there is regular seismic activity in the area, so the Theran Eruption could have coincided with other earthquakes to turn the Nile red.

ii. A plague of frogs

The frogs would of course leave the river if the CO_2 level went too high as they would not be able to exist under those circumstances in the river. Then thousands of frogs and toads would have swarmed over Pharaohs palaces and temples causing great panic to the priests and officials.

iii. A plague of locusts

The ash cloud could have descended onto some of the crops in the fields encouraging swarms of locusts to take to the air to look for fresh fields of wheat. Whilst traversing to ash free areas the locusts would have gone through the Egyptian Cities of Avaris, Pi Ramses and other cities causing much panic and annoyance to Pharaoh and his subjects.

iv. A plague of murrain afflicted the cattle

The flies would have carried diseases which would afflict the cattle. Also there would be no decent water for the cattle to drink further weakening the cattle making them more susceptible to insect borne diseases.

v. A plague of hail (ice outside and molten rock in the centre)

This actually sounds like a volcanic ejection into the atmosphere of spheroids and or tephra which may (in a large volcanic eruption) have been blasted by the huge pressures of an explosive volcanic eruption (such as Thera) into the upper atmosphere where the molten hot spheroids (small blobs of molten lava) would have become coated with ice due to condensation effects and then would have dropped back to earth (Egypt is only 700 miles from Santorini and well within the ash plume range) as ice covered hail with a red hot interior, just as the bible said it was. This of course could have altered the chemical makeup of the Nile making it seem blood red and probably killed or injured many more of the cattle in the fields.

Lumps of pumice stone from the Theran eruption have been found in Egypt.

vi. A plague of boils

This could have been caused by the plague of flies or lack of vitamins and proteins as all the cattle had died or was inedible and all the Egyptians would have had to eat would have been some dried fruits, nuts and stored grains.

vii. A plague of darkness

When the ash cloud reached Egypt about a day or so after the initial eruption the sky would have darkened quite quickly and would have blotted out the sun for some days, therefore causing the affect of darkness during daytime. The Egyptians used to a baking hot cloudless blue sky would not have know what had hit them and would have been truly terrified.

viii. A plague of lice

Similar to flies and locusts, lice would thrive in these conditions of rotting animals and unwashed Egyptians and would have been annoying to the Egyptians.

ix. A plague of vermin

Thousands of rats and mice would have been attracted to all the rotting carcasses of all the dead cattle in the fields and once all these carcases had been devoured the vermin would have headed into the houses, palaces and temples of the Egyptians no doubt terrifying all the ladies, priestesses and princesses in Thebes, Luxor and Pi Ramses.

x. A plague of pestilence

All these insects, dead animals, lice and frogs would no doubt have carried many diseases which would then have struck down many of the Egyptians in their cities.

xi.   The death of the firstborn

This is an interesting phenomenon, based on increasing amounts of CO_2 leaking through seismic vents into the river Nile. There has been several cases of this sort of thing happening in modern times. There is a lake in Cameroon (West Africa) where CO_2 levels rose in the lake turning the lake blood red (see above), then one night the CO_2 levels got so high that the CO_2 (carbon dioxide) bubbled into the atmosphere and rolled (invisibly) down beside the lake and through several villages, the next morning many people men, women and children were found dead due to suffocation by carbon dioxide poisoning, the completely invisible, colourless and odourless gas had simply bubbled out of the lake and rolled down away from the lake as this gas is heavier than air over the countryside killing all humans and animals in its path.

Why, you ask though would it have killed all the Egyptian firstborn? Good question, however the answer is very simply explained. In Egypt in Biblical times the firstborn (males) of the household slept on very low beds, other members of the family slept up on the roof or in higher beds or higher rooms. It was a mark of respect to let the firstborn son sleep low down (like Pharaohs bed), therefore when the CO_2 gas rolled off the Nile it killed all the firstborn sons and the rest of the family who were sleeping on higher beds escaped death by suffocation.

Whether the volcano erupted due to gods intervention to punish the evil Pharaoh and his people for not allowing the Children of Israel to leave Egypt or whether it was a natural phenomenon is a subject for another book and not open for discussion at this juncture although it must be said that the Biblical writers did seem to capture much about this mega explosive volcanic eruption that you could imagine being described by CNN in downtown Cairo watching events actually happening.

C. The Eruption of Mount Vesuvius in 79 ACE

Nearly everyone knows of the cataclysmic eruption which destroyed the cities of Haracleum and Pompeii in Southern Italy in the first century ACE. I will not dwell too much on the implications of this terrible destructive eruption, we have all seen "Up Pompeii" with Frankie Howard, some might think after seeing this comedy that it was a good thing that Pompeii was destroyed. In reality the speed of the eruption caught many inhabitants by surprise and many residents were asphyxiated by the fumes and gases and then their bodies preserved in the layers of ash which in places were 10 meters deep.

Today this sleeping giant is surrounded by about 3.5 million people all living within 20 – 30 miles from the crater, if and when Vesuvius blows again, and she will, many hundreds of thousands of people will lose their lives. The Italian Government is trying to re house those residents living very close to the volcano, however many hundreds of thousands of people are still living in Vesuvius shadow today.

D. The Events of 534 – 535 ACE

Major events around the world at this time are attributed to some major volcanic activity, however volcanologists cannot agree what or where blew its top. However there were major world changing events brought about by these mysterious eruptions, some scientists claim that Krakatoa in the Java area blew its top off, as is corroborated by the Javanese Book of Kings, however The Book of Kings maintains that it blew in 416 ACE and not 535. Other scientists claim that it was the Volcano Rabaul in Papua New Guinea, we may never know the real culprit for the following cataclysmic effects across the globe.

Documentary evidence of the 534 – 536 extreme events

The Byzantine historian Procopius reported in his tome, "History of the Wars, Book III and IV: The Vandalic War," that "during this year a most dread portent took place. For the sun gave forth light without brightness......and it seemed exceedingly like the sun in eclipse, for the beams it shed were not clear."

The Annals of Ulster recorded: "A failure of bread in the year 536 AD."

The Annals of Inisfallen wrote: "A failure of bread from the years 536 – 539 AD."

China: Low temperatures, snow even during the summer months. It is reported that snow in August postponed the gathering in of the harvests for some time.

Mid East, China & Europe: reports from these areas of "a dense dry fog."

Peru: A severe drought in Peru affected the Moche civilization.

i. **The Justinian Plagues across the Roman Empire**

A major outbreak of bubonic plague swept the Roman World in 534 ACE, we think that the famine was brought on by a volcanic freezing winter and a summer without sun destroyed major crops and food sources in the Eastern Empire. Due to the decaying cattle and general unhealthyness of the times an outbreak of plague ravaged the Eastern Empire and then spread to the remnants of the Western Empire and into Gaul, Spain and Italy. There are reports that the Romano British who inhabited the Western part of the British Isles at that time, due to their continued trade with the Empire caught the plague, however it is stated that the Anglo Saxon settlements only a few miles away were unscathed due to the lack of communication between the Roman Britons and The Saxons.

Another field of thought is that due to crop failures in Egypt and the Middle East generally, Roman Merchants and traders pushed higher up to the source of the Nile towards Africa and traded with the African Tribes around Lake Victoria which unfortunately for them (and many others) is still the source of many of the deadliest diseases known to man including ebola, bubonic plague, even possibly AIDS. It is thought that some of these merchants or their slaves caught plague like symptoms and carried it back to Constantinople from where it spread out across the late Roman Empire.

It is commented on by various Roman writers that as many as 30 % of the population of the Roman World died during this awful plague.

The repercussions of this decimation of the population of the late Roman Empire were like ripples spreading out after a pebble had been thrown into a pool. Justinian and his generals Belisarius and Narses had re conquered much of Italy from the Visigoths after a protracted series of campaigns which eventually was to cost the Byzantine treasury 300,000 pieces of gold, a phenomenal amount for those days and had done some damage to the barbarian tribes settled in Northern Italy and had stabilised the frontier. These wonderful generals had also re captured Libya and the rest of North Africa from the Vandals who returned to Spain to lick their wounds. As happens so many times in history just when you start to relax as Justinian and Theodora his beautiful "actress" wife must have started to do, all hell broke lose again, wars erupted in Persia with the Sassanid Empire and during all these struggles in the late 530's plague struck the empire, in my opinion fatally weakening its internal well of man power and causing massive economic hardships, this together with Justinian overstretching the empire's resources in re conquering the Western Roman Empire caused after his death in 568 ACE a partial collapse of the newly won empire in the West to the Lombard's, Franks and other barbarian tribes.

Italy itself, having been fought over for 100 years was practically depopulated by these incessant wars and fell prey to many invaders and fundamentally influenced the way Italy would later grow up as a series of city states and other areas controlled by foreign empires. As we shall read about next, the plagues and economic and manpower collapses practically allowed the Arab invaders from the deserts to walk over much of the weakly held Roman Empire, which perhaps, if the volcanic eruptions had not forced the Roman merchants to pursue their hunt for grain into the Lake Victoria region would have enabled a stronger Byzantine Empire to fend off some or all of the Arab attacks.

ii. **The Arabic Invasion of the Middle East in the seventh Century ACE**

The Byzantine Empire (Eastern Roman Empire) was so weakened by this terrible plague that it didn't ultimately have the manpower some years later to withstand the up surging of Islam from Arabia and the Hejaz across the Middle East, there was an important battle near the Beka Valley in Lebanon where the Byzantine army hardly even put up a fight and simply fled when the Islamic armies advanced. Modern day Israel, Lebanon, Jordan, Syria, Egypt, Libya, Morocco, Tunisia, Algeria, Spain and much of Turkey fell to the Arab armies due much to the numerical weaknesses of the Byzantine troops and the economic weakness of the Byzantine Empire.

In Egypt, Libya, Tunisia and Morocco the same thing happened, an enfeebled giant lay prostrate as the vigorous armies of Islam marched over the land conquering all before them. What would have been the difference had the plague caused by the volcano not struck down the Byzantine Armies and they had a Belisarius or Julian (the Apostate) or a Caesar or Trajan to lead them, in all probability they would have hurled the Arab armies back into the deserts of Saudi Arabia from whence they came. The whole history of Europe and the world would have been very different today had this plague not infected the Empire at that critical moment.

iii. **Famines in Egypt, India and the Far East**

Reports from Ireland to Japan state that there was in effect a volcanic winter across much of the earth for many months which changed the course of the monsoon rains and affected other weather patterns so that terrible famines occurred, no rains for months on end and terrible scenes of starvation on death on an epic scale resulted from this calamity.

iv. **Decline of Teotihuacan, in Mesoamerica**

This demise of this huge city is Central America is also linked to the droughts and famines of the period incorporating 535 – 536 ACE. Some say this city was the site of one of the Atlantean Cities dating back to before the last Ice Age, the jury is still out on this concept though.

v. **Decline of the Avars and migrations of Mongolian Tribes**

It has been postulated that a large warlike Hungarian Tribe, the Avars declined in power due to climactic changes in Eastern Europe and that mass Migration of the Mongolian Hoards occurred at this time as well.

E. Mega Eruptions affecting life in the Middle Ages

There were reports from chroniclers of many strange atmospheric and seismic events in the Middle Ages in Medieval Europe that some scientists think indicate major volcanic activity in Iceland and other parts of the world at this period in time.

As I will discuss in later chapters, this is the time of the ending of the Medieval Warm Period and the onset of The Little Ice Age in Europe and some scientists are adamant that the smoking gun for this cooling down period may have been some mega eruptions in Iceland or The Far East.

F. The collapse of the Mayan Civilization 11th - 12th Century ACE

Many historians and archaeologists have postulated the reasons for the decline and fall of the Mayan civilization in the early middle ages, some archaeologists have stated that internecine wars between the Mayans and the up and coming Aztecs and other local tribes was to blame, others think that plagues and famine's caused the de population of the Mayan cities.

My theory is that a volcanic eruption such as Hekla's H1 event in 1104 ACE which caused much misery in Iceland and Europe and actually caused some migration off Iceland and back to the mainland could have been the culprit. It is said that after this eruption that the whole of Europe was familiar with the name of Hekla, my theory is that, like Laki in 1783 the size of the eruption and the large scale areas of land and sea affected by the ash clouds possibly caused the movement of weather patterns such as the monsoons and El Nino, which in turn caused sustained droughts and drier weather to occur in the area of the Maya's and associated tribes, which caused them to abandon their cities as they could not keep up the infrastructure and the hierarchical nature of the priest kings and therefore after an orgy of human sacrifices and inter tribe warfare their civilization broke down and the survivors drifted back into the jungles of Central America where to this day their descendents still live.

I do understand that without weather circulation patterns for the 11th & 12th centuries it will be difficult to prove this hypothesis, however I do feel that this synergy of volcanic eruption, changing weather patterns and major droughts has occurred in the past and will occur again in the future.

G. The Great Famine 1315 – 1317

This disaster struck Europe more or less without warning in 1315, before this period there had been a period called the Medieval Warming Period when vines and agriculture expanded in Europe and population levels reached as high as early Nineteenth Century levels. The Great Famine started in the spring of 1315, when the spring was unusually cool and it rained and rained for months, flattening crops and not allowing grain and wheat to grow properly. Food prices doubled between spring and summer, there was no fodder to feed livestock and no salt to cure meat and bacon. By summer many of the lower classes started to starve as they could not afford the high price of grain, the peasants were scavenging in woods and forests for roots, berries and acorns and bark to find some nourishment. When Edward II stopped at St Albans on 15th August 1315, no bread could be found to give him, which shows how bad things had become.

In 1316 it continued to rain and by now the weakened population was really beginning to suffer, draught animals, horses, donkeys, milk cows were all slaughtered and seed corn was used to eat. Some children were driven out to fend for themselves, this is where the story "Hansel & Gretel" comes from, some people even turned to cannibalism in order to survive, hundreds of thousands of peasants started to die of hunger. In 1317 the height of the Famine was reached with thousands succumbing to starvation, bronchitis, tuberculosis and many other debilitating diseases. It is estimated that up to 25 % of the population died of hunger and other related diseases. Later in 1317 normal weather started to return, however it was not until 1325 that matters started to return to normal.

The consequences of this great famine were threefold, firstly it undermined the church's power, as no amount of prayers seemed to change the weather patterns, secondly Europe became a more violent place where murder, rape and all manner of crimes were commonplace and feudal society seemed to break down. Thirdly the earthly Governments of Europe failed to do anything about the famine. The famine weakened Europe considerably and when the Black Death struck in 1338, the enfeebled population succumbed to the Black Death in much greater numbers than normally, due to the weakened state of most of the population.

As a matter of interest some scientists who have recently been studying the Black Death do not now think it was caused by bubonic plague carried on rats by fleas from the Levant to Genoa in Italy, they think instead after various examinations of the skeletons of infected persons that it may have been a form of the ebola virus which in effect causes the internal organs to melt and dissolve which obviously causes a pretty horrible death and there is no cure to this disease to this day. What is further very interesting is that some people have a natural vaccine to this disease and are not affected by it, no one knows why though.

There is some incidental evidence from various chronicles and annals that some Icelandic or other volcanism in the early part of the fourteenth century could have affected weather patterns considerably and caused the violent changeable weather which caused the Great Famine of 1315 – 1317. It is my belief that in time a major culprit will be seen to be one of the big Icelandic Volcanoes, this episode again shows how helpless and vulnerable we are to really large natural effects on the earth, although with today's technology and the ability to ship large amounts of food fairly swiftly around the world the effects of The Little Ice Age, were they to be repeated today, would not be nearly as severe as in the 14th century. In my opinion we do however have to look carefully back in time at all these different climactic events which have altered the evolution and growth or decline of civilizations to learn how to cope should similar situations arise in the future. Our governments have to plan for a time when for up to three years no food can be grown across Europe, they had better start building mountains of canned foods rather than lakes of wine and butter mountains.

H. The Little Ice Age 1500 – 1816

There are many interesting features about the Little Ice Age (LIA), there are arguments over when the Little Ice Age actually started, some scientists state that it was in 1250 ACE when the Atlantic Pack Ice started to increase in size, 1300 was when warm summers stopped becoming dependable in Northern Europe, 1315 – 1317 for the start of the two years of the great rains and the Great Famine, 1550 for the beginning of the worldwide expansion of glaciers and 1650 for the first climatic minimum. For arguments sake I have postulated that this is a period after about 1500 ACE when winters started to get much colder and many rivers in Europe froze over regularly during the winter months, many famous Dutch paintings of the early and mid seventeenth century by Breughel the Elder and the younger depict winter ice skating scenes on frozen rivers and lakes and these sorts of winter scenes were never depicted before this time. We also have evidence that many European Alpine Glaciers were on the move and it appears that quite a few towns and villages were actually destroyed by advancing glaciers during this period. As late as 1930 the French Government commissioned a report on the advancing glaciers and how it would affect France.

The change in temperature's affected agriculture, vines could no longer be grown in the UK and most of Northern Europe, where many crops could not now be grown and many farmers turned to hay and fodder production, to keep herds of cattle and cows and Dutch cheeses started to be popular, the Dutch in particular learned how to cope with the changing weather and their technical abilities in making windmills and draining water from swampy, marshy areas enabled them to increase food production during this period. Many root crops were then sowed including parsnips, swedes, carrots, potatoes, beans, peas and other similar crops and many advances in agriculture and allowing fallow lands followed. This cold weather changed people's house building techniques and encouraged new shipbuilding techniques and the Portuguese, Spanish, Dutch and British all set sail to carve out trading empires across the world partly due to the Little Ice Age and its changing of the existing economic systems.

In Iceland itself the sea ice actually closed in around the coasts making it almost impossible for fishermen to put to sea in their boats. The freezing weather, blizzards and the inability to grow crops led to mass migration away from Iceland back to Europe, it is estimated that as many as 50 % of the population of Iceland emigrated due to the harsh inhospitable conditions encountered during these years. This was on top of regular volcanic eruptions making things even worse for the poor Icelanders.

Greenland also about his time became so inhospitable that all the Norse settlements vanished at this time and only the Inuit (Eskimos) were able to eke a living out of this barren frozen land.

There is a theory that between 1500 and 1816 there were some pretty massive volcanic eruptions in Indonesia and the Pacific Islands of New Britain and The Solomon Islands and some of these eruptions changed weather patterns and may have caused a cooling of the world for some years. I am not saying that the volcanic eruptions were the only cause of the Little Ice Age, the El Niño effect could also have come into play, however they may have tipped the scales and caused the temporary movement of the Atlantic Conveyor, (the warm ocean currents from the Caribbean to Norway) or they may have been accentuated by The Maunder Minimum which was the name given for the low level of sunspot activity which corresponded with the Little Ice Age. Before LIA (Little Ice Age) high pressure over the Azores and the low pressure over Iceland which was the norm in weather patterns, but somehow this became reversed in the sixteenth, seventeenth and eighteenth centuries and therefore caused easterly winds to blow across Arctic Russia and across mainland Europe for most of the year bringing excessively cold weather to much of Europe for many years. Towards the end of the Eighteenth Century we must not forget Laki's eruptions which would definitely have cooled down the temperatures for several years which kept the cooling effect going longer than would otherwise have happened.

It is my surmise that the combination of a strong El Nino effect twinned with some major eruptions in the Pacific area, including the eruptions of Mount Billy Mitchell, 1580, Mount Parker, 1641, Long Island (Papua New Guinea) and Huaynaputina in 1600, pushed Europe and North America into this major cold snap for about 250 years. Some of the volcanic eruptions in the Pacific were of VEI 5 and 6 levels with tens of square kilometres of eruptive materials being blown into the upper atmosphere and the sulphuric acid droplets would have reflected sunlight for many years which caused a huge cooling effect on the whole earth. The fact that until early Victorian times the Thames froze over regularly to such an extent that frost fairs were actually held on the Thames as the ice was many feet thick and could support fairs, supports this theory.

I. Volcanic eruption which caused the final downfall of Constantinople 1453

A huge undersea volcano Kuwae near Hawaii blew apart in about 1453 ACE causing probably huge tsunami's and major devastation in the Pacific, which we know little or nothing about currently.

It has been postulated that changing weather patterns in 1453 caused crop failures that weakened the defenders of Constantinople and eventually enabled the Ottoman Turks to capture the city thereby extinguishing the last spark of the old Roman Empire.

J. Tambora super volcanic eruption of 1815 – 1816

This was a major earth changing event that caused many changes in the social fabric of society both in Europe and America. This eruption caused the famous "year without a summer" in Europe and caused many freak lighting storms across Europe, it could have contributed to the incessant rains which hampered Napoleon's last stand at Waterloo as his troops were in the lower lying region of the fields of Waterloo and would have to have marched uphill through boggy wet soil to reach the English and their allies at the tops of the hills.

This was the year that Lord Byron and Shelly with Mary Shelly and Byron's mistress spent the season at a castle on Lake Constance near Geneva and there were heavy rain storms and massive thunder and lightning storms due to the volcanic ash in the air from Tambora, these storms and frightening aspects are said to have given Mary Shelly the inspiration to write the bestseller Frankenstein whilst staying at the chateau.

In the USA the winter of 1816 – 1817 was one of the coldest ever recorded in new England and surrounding states, the frost lay on the ground for months and months and many people and animals froze to death in that year.

Due to the lack of a summer because of the ash cloud from Tambora covering the globe, crops failed across the world and there was much starvation and misery that year across the globe. One must not forget that this was not a full blown super volcanic eruption, like the one that nearly wiped out humanity 75,000 years ago, this was little more than the rumblings and tossing and turning of a sleeping giant and we should be aware of the very real risks to humanity if one of these massive super volcanoes actually blows up as they have done regularly in history.

K. Krakatoa Java 1885

This blast could be heard 3000 miles away and was said to be have been the loudest noise ever to have been heard on the Earth in historic times. Massive tsunami's and thick layers of ash devastated hundreds of miles of coastline, it would have been like the 2004 Boxing Day tsunami with a huge volcano blast and ash fall on top. Many hundreds of thousands of Javanese and Indonesians are thought to have perished in this monster eruption. Red sunsets were seen in the UK for months after the eruption as ash was trapped in the upper atmosphere for some time.

I feel that in the fullness of time it will be recognised by archaeologists and historians that many large eruptions in Iceland and elsewhere in the world have caused either collapses of civilizations and/or a deviation to the path of human historical evolution that has radically altered our historical view of the world that we live in.

As an example who knows why the winter of 1812 was so harsh that the French when they had defeated the Russians at the battle of Borodino and then actually reached Moscow and the Russians set fire to Moscow they had to retreat through one of the most bitterest winters on record in which about 90 % of Napoleons army froze to death.

Similarly in 1941 when Hitler's Wermacht was so close to Moscow that the advance troops were bringing back books of tram tickets from dispensing machines on the outskirts of Moscow, when the barometer suddenly fell to about − 40 deg C and it was so cold that diesel froze in the panzer tank engines and fires had to be lit under the tanks to keep the diesel from solidifying, gun barrels froze so that shells could not be inserted into them and German Soldiers who had to go and relieve themselves froze to death in seconds. It was one of the coldest winters in living memory, yet it saved the Russians from being overrun by the Germans, as the Russians were used to operating normally in these extreme cold weather conditions, maybe caused again by eruptions somewhere in the world that had changed weather patterns for a few years making them a few degrees colder due to the sulfurdioxide particles reflecting back the sunlight causing the earth's temperature to drop further than it otherwise would.

If this is found to be the case then once again it can be deduced that eruptive volcanoes have dramatically changed the evolution of historical life on earth. There are many instances of extremely cold periods such as the mini ice age of the Tudor and Stuart eras when the Thames regularly froze over and frost fairs were held on the Thames, and the warming period of the early middle ages when vines were grown across the UK and many monasteries in the UK were producing fine wines for their own consumption and for export and sale. I do feel that in the fullness of time that it will be shown that changing weather patterns caused by volcanoes contributed to the many changes of global weather and climate patterns, just a few of which I have mentioned above.

So what are we all going to do about it? How are we going to carry on with our day to day lives, however with the thought that in the backs of our minds a day will come when somewhere in Iceland, Italy, USA or the Far East a large volcano blows its top and we have to make some preparations? Firstly, let's discuss what will actually happen out "in the field" as "Q" from the James Bond Movies always berates Bond for destroying or damaging his expensive equipment and Bond always says, "it's pretty tough out there in the field Q." It will also be tough for us in the field after a large volcanic eruption.

The UK Governments reaction

We all pay a lot of tax and now VAT has gone up to 20 %, Capital Gains Tax has nearly doubled to 28 %, so what preparations has the UK (or any other of the major world governments) done to prepare for this eventuality. I have written to H M Government to actually ask this question (no reply to date at time of publishing), my guestimate is that nothing has been done, no stockpiles of tinned foods, no underground reservoirs of potable drinking water, no emergency stockpiles of oil, baby food, rugs, blankets, dental equipment, medical equipment, anti biotics, animal feed, the list could go on for a long time with the same dismal answer, nothing is being or has been done in preparation for the forthcoming disasters.

In the 1950's when the UK was locked with the USA in a "Cold War" with the Soviet Union some provision was made, although laughable by today's standards, a serious voice announced the "Protect and Survive" broadcast about how to survive a nuclear attack from Russia, it proudly stated "if you hear the sirens you will have about 5 minutes to make your final preparations", what were these lavish preparations that we were all going to make in order to be safe and sound. We were all going to get a couple of mattresses and the odd blanket and pillow and huddle together with a bottle of mineral water under the dining room table, of course that would protect all 50 million Brits from the 30 missiles each of which was about 40 times stronger than the Hiroshima and Nagasaki atomic bombs, oh sorry I forgot, we were to put sheets or blankets over our windows to protect us from the nuclear flash and hurricane force winds and days of invisible radioactive fallout which was going to be blown by the wind all over the place.

Whilst we were all singing "there will be blue birds over the White Cliffs of Dover" and huddling for 3 years under our dining room tables, where were the top government officials, town hall staff, the military and other "vital services" going to be hidden, was it under their tables in their dining rooms, bloody well not is the sensible answer, H M Government had spent untold millions in the 1950's and 1960's under both Labour and Conservative Governments on building some pretty lavish self contained underground cities (well large concrete bunkers actually) with filtered air, stored safe water, tinned food, beds, kitchens, heating, medical supplies and all the other items from the above list that they didn't have for the general population. They could then sit out the three years of radioactivity playing monopoly, canasta, bridge and watching re runs of Big Brother II whilst most of the population died of radiation, sounds fair to me.

Seriously though, it would be financially practically impossible for the Government to have built underground cities for the whole population, we just didn't have the time, or the resources to do this and the Government will quickly tell everyone that someone has to pick up the pieces and carry on with the business of running the country after the three years is up. The USA had the same philosophy as us Brits and hollowed out several large mountains in the Mid West to use as nuclear proof control centres. Funnily enough, the Russians actually did construct some large underground bunkers for many city dwellers, although it could be said that it wasn't a very sensible place to build the bunkers right under where the British and American nuclear missiles were going to fall, somewhere on the outskirts of cities would probably have been more sensible. In Switzerland under Federal Swiss Law every property constructed since 1962 actually does have to have a nuclear bunker built into it, in the basement. The Swiss also have communal nuclear shelters on the outskirts of many of their towns and cities, those cleaver Gnomes of Zurich wouldn't be caught dead under the dining room tables like us Brits, we however are manlier than the Swiss and like to face Armageddon properly ensconced under our tables like real men and women.

The point I am trying to make here is that when CNN or Sky News show you the ash plume on TV there is no nice man from the Local Authority who is going to start driving around your streets with a whistle saying follow me into the "Government Ash Fallout Shelters" down by Corporation Street as, nothing at all has yet been planned for by the Government. We are going to have to make our own provision and look after ourselves. Luckily for us the volcanoes are not radioactive and they do not land on Manchester with a 30 mega ton warhead. Just soft white or grey (or invisible) ash will start to fall, maybe some spheroids, a little sulphuric acid rain, killer fogs etc, but whatever you do don't forget to hide under the dining room table for the first half an hour (just to get used to things) and to give the neighbours a laugh.

What will the bulk of the population do when newscasts start requesting that the public stay indoors for 3 or 4 days at a time, possibly without power, water or sewage systems working, it will be too late to do much at that point in time, preparations have to be made well beforehand. I don't know how severe the problem would be, it could be similar to the EJ eruption where only a few flights are disrupted for a week or so and the price of Turkish (air freighted) strawberries goes up by £ 1 per pannet, or it could be a Katla, Laki, Yellowstone or Tambora event in which case there could be up to 2 years worth of disruptions to normal life, without power, water, sewage, food, petrol or medicines to contend with, so what do we do???

Firstly we have to try and work out exactly how bad the situation is going to be and for how long, if all that's going to happen is that a few Easyjet and Ryanair flights are cancelled for a couple of weeks, whilst that is inconvenient and annoying if you were actually planning on going away on business or holiday, it's frustrating, however it's not the end of the world and your life goes on, you can still go to Tesco and Asda and BP and buy what you want, life carries on as per normal.

However lets set up the scenario say, that Laki erupts at a level VEI 7 (Volcanic Explosivity Index) eruption (1 being mild and 10 being cataclysmic) what would happen, firstly the huge ash cloud would ground all air flights over the Atlantic, cutting off Europe from the USA (apart from ships) for maybe several months maybe a year, the first casualties would be families on holiday trapped in the USA and Caribbean and South America who wouldn't be able to return to their families and jobs for weeks or months, the recent ash cloud was tiny compared to what Laki could throw up and nearly every office and family had tales of stranded friends and relatives who in some cases took 3 – 5 weeks to get back to the UK, there were buses laid on to Spain, chaos at the Channel Ports as Brits tried to get back to the UK, people spending thousands of pounds on taxis and hire cars to get themselves from Italy to Calais. If a key member of a company is trapped abroad for months that would put great pressures on that company and the other staff left holding the fort, although with laptops and Blackberries' these days people can stay in touch, files, meetings, conferences, purchasing decisions will all be put on hold maybe indefinitely until the key member of staff returns to their job. Of course families will be devastated if mum, dad or sisters or brothers are trapped abroad, especially if it's the bread winner who cannot return home for some time.

The next thing to hit home will be any goods, medicines, vaccines, machinery parts for power stations, rail networks, mobile phone networks, and factories or the military that need to be urgently airlifted from the continent or the USA back to the UK. The only alternative would be to put them on rail or road transport and get them under the Channel Tunnel, however we all saw how quickly that route backed up immediately the planes stopped flying, urgent parts and life saving equipment for the NHS and phone companies could be sitting in container / trailer parks near Calais for weeks and months maybe as the ports and Channel Tunnel clog up. You cannot suddenly organise hundreds of new container ships and ferries to travel between

Calais and Dover or Boulogne to Folkestone at the drop of a hat, everything takes time and re scheduling. All it needs is a part to fail on one of the main electrical sub stations around London for a widespread power cut in the capital to be triggered, this could even have a knock on effect for London Underground as they recently closed their own power station at Lots Road and are now powered by the National Grid, so a small piece of equipment failure and the inability to get the correct part quickly from abroad would mean that potentially a swathe of London and the Home Counties could face intermittent black outs which could also seize up the over ground rail networks, South West Trains, Silverlink and South East Trains and the Underground could all grind to a halt. If the so called "Laki Fog" also descended on the UK then difficult situations would become dangerous situations. If this sulphurous fog was severe the oil tankers would not be able to put into Milford Haven and the other tanker depots around the UK, the super tankers which take 7 miles to slow down and stop would have to anchor out to sea and wait until the fog cleared. If the fog lasted about eight weeks with a ridge of high pressure over the UK and light winds (as happened in 1783 – 1784) then very soon petrol rationing would come into place, for those who can remember the oil crisis of 1973, petrol was rationed to about two gallons per fill up and queues miles long quickly developed at all petrol stations. Then power stations that ran on oil would have to cut off the power for part of the day, possibly evenings and weekends at first to ensure that industry, the rail network and hospitals carried on receiving power supplies. The UK, unlike France and Germany has only about 1 week's supply of strategic oil for petrol and diesel after that it's all stop for petrol and diesel and then it's back to the horse and cart for everyone.

In a harsh winter there could well be serious suffering for the disabled, elderly and infirm citizens as the heating and hot water and power supplies were interrupted on a regular basis. Without oil, supermarkets would not be able to get food from container ports (if any of the container ships could reach the ports) and the Government would probably step in to ensure that what meagre supplies of oil were left was given to the road haulage companies to keep food, milk and other essential supplies moving.

My hypothesis is that a volcano like Katla or Laki would temporarily, for maybe 3 - 6 months to a year have the symptoms of a super volcano, as we are living fairly near the epicentre of any eruption and that eruption (as in the past) could rumble on for 1 to 2 years. My theory is that apart from having the possibility of a "Laki killer fog" hanging around for several weeks in the summer (or Autumn) which would mean that no one would be able to leave their flats or houses safely, that combined with this we could also have several seasons of widespread crop failures and freezing winters without air travel for up to 2 years.

What can the Individual and or family do about it

Advice has to be general, some people live in bedsits, some in hotels (on benefits), some live in small studio flats, penthouse flats, terraced houses, mansions, castles or palaces, therefore you will have to make proportionate judgements depending on how many people you are living with, do you want to include other members of your family, all your street, everyone in your block of flats, etc, etc. I cannot help you decide whom you will want to assist and whom you don't help.

However I would make one valid point, if you do make preparations, store some food, water, medicines and gas cylinders and then you go down the local pub and tell everyone there what you have done, when the ash starts falling in your street you can expect many "new friends" to arrive at your flat / house and then you might realise that you haven't enough water, food, blankets to go round and then all hell will break lose as you try and control the situation and people start to become desperate. I cannot really advise you as to how many people you tell about your preparations, it's going to be your call, I would like to save every one of the estimated 62,000,000 residents of the UK, however unless the Government starts massive bunker building and stockpiling from tomorrow onwards so that everyone can be fed and watered for between 1 and 2 years it's not going to happen.

So, let's say you are a family of five, two adults and two children and your wife's mother (never forget your mother in law, or else), so you have five mouths to feed and water for six months (it would have to be pretty catastrophic to go on more than six months in my estimation). Each person would need at least two cans of food per day and one litre of water per day, that's 70 cans of food per week and 35 litres of water per week. I do however think that the Government will be able to organise some basic food distribution after about three weeks to one month, as other countries (not yet affected by the ash clouds) ship emergency supplies over to the UK. However as I previously stated, whether H M Government can fully feed and get enough uncontaminated water for 62 million souls is asking for some sort of miracle.

However for our hypothesis, I am supposing that some sort of water stand pipes will be put up or water tankers will come round the streets, bread and canned food distribution centres, or perhaps rationing books like during WWII will be issued, so my theories do suppose that some minimal help will arrive from the authorities. However whenever I look at news broadcasts from Haiti, Ethiopia or Bangladesh it always seems that there is a massive crush and fighting involved in trying to get the odd packet of rice or bottle of water, also there always seems to be massive corruption in that food stuffs supplied by charities ends up being sold openly in towns and villages in the emergency areas and I for one do not want to just sit and wait for some meagre provisions that might eventually be distributed in some fighting scrum with the rest of the street barging in to the relief centre.

What will happen to elderly and disabled and blind people living on their own, hospitals, old people's homes, mental institutions, battered and single women's refuge's, sheltered housing communes. This will be a nightmare for the Government to organise a relief effort on this scale, the logistics and forward planning, even if they started now are horrendous to think about. We have to imagine that a huge ash cloud has spread over much of the UK, Europe, parts of Russia and across the Mediterranean as well as most of the Eastern Seaboard of both the USA and Canada, all flights to the UK, USA,

Europe and Russia are grounded. The nearest open airport might be Tunis or Casablanca in North Africa, hundreds of thousands of Brits will be stranded all over the world for months on end and many Americans will be stranded over here. If the killer fogs also appear (which is likely as many of these volcanoes are under glaciers which then make the killer chemical cocktail of the poison haze by reacting with the glacial water when the volcano erupts) then the oil tankers won't be able to dock and they will have to anchor somewhere off the Bay of Biscay or even near Gibraltar to be safe. People will start panic buying and very soon most supermarkets will soon become empty and the supermarkets will not be able to restock the shelves as petrol supplies and diesel supplies run out. Everyone will also try and fill up their cars with petrol, therefore the petrol stations within about a week will run out of petrol. The reservoirs may become poisoned by the ash and chemicals in the gas cloud and fogs, although it is not clear as to at what point the open water reservoirs could become undrinkable, there is the possibility though that the supply becomes contaminated and then the government will send out water tankers which will cruise the streets or stop at scheduled stops in each neighbourhood so that people can fill up containers with drinking water.

Most of the crops in the fields will be covered in a thick ash which would render them inedible, so wheat, barley, peas, and surface crops will immediately be affected, with regards to root crops like potatoes carrots, onions etc, subject to the ground being tested for acidity and chemical toxicity by government inspectors it might be possible to still harvest some of the root crops. However farmers and farm labourers would have to wear gas masks whilst working in the fields to avoid death from the killer fogs.

If any major component fails in the National Grid electricity supply or the weight of ash brings down the pylons or cables then many areas would have blackouts, if a major component failed (and let's not forget that most of our National Grid system is about 40 years old) and that part had to be obtained from Germany, France or USA, how would that part get to the UK, the Channel Ports would be massively clogged with container ships and ferries trying to get Brits and cargo's back to the UK, Eurostar services would be sold out for months ahead and there might be no fuel to drive the part to where it's needed. I am in the residential property management business and look after many blocks of flats in the London area, recently I had to order a part for a lift from Finland and with everything working well it took about 6 weeks to arrive in the UK and be shipped to London, who knows how long it would take to get urgent replacement parts to the UK if a major ash cloud hits us. What about urgent hospital equipment to keep operating theatres going, flu vaccines and other specialist medicines, parts for factories, engineering, railways, nuclear power stations, wind turbines, telecoms companies, the list is endless and as stated before even now with everything running well things take ages to arrive, what will happen when nothing is running.

If there are power cuts then the over ground trains and underground system will grind to a halt, shops and offices and factories will have no power and will close, lifts, electric doors to blocks of flats and offices will fail and basically our whole way of life as we know it will grind to a halt whilst the volcano continues to erupt. I am not stating that everyone reading this book will cope 100 % with all the problems that will be encountered during this potential hazardous time to come, however if you follow my ideas you and your family will have more of a fighting chance that those of us who don't bother to take any precautions and put their faith in the Government and the EU.

So what can be done today to start preparing for this forthcoming nightmare scenario?

What will I need to bring into my home?
1. Food

Let's assume that the average family will be four persons, all my calculations are being done to cover four people so if there are only two of you then you halve the figures I am showing, if there is only one of you then you only have to work ·on 25 % of my figures. This will not be an exact science as a large rugby player will probably eat a lot more than a three year old, so at best these figures must be taken as a guideline and not as hard actual figures. Careful planning will need to be done about how and when to eat the food stored up, some foods will require cooking, boiling, grilling etc, and you will have to carefully work out portion controls to ensure that everyone gets their fair share of the available food, on the basis that it must last for many months possibly. All this should be worked out well in advance of the arrival of the volcanic ash cloud.

As I stated earlier, the food that you will have to store will have to be food which is either canned, freeze dried, powdered, or preserved in some way, foods like powdered potato, dried fruits would be useful, as they will contain fibre and some vitamins and proteins. So basically what you need to look for at the supermarket is mostly preserved foods including perhaps some chocolate and sweets (which actually also will last a long time). Do not choose any fresh fruit, meats or bread etc as this will all go off without refrigeration within days and is unsuitable (without power) for long term storage.

Another idea is to create an 18th Century ships apple barrel, on board all of H M frigate's and sloops in the late 18th C and all of the 19th C, ships had barrels with apples piled inside in rows, probably with a little straw in the barrels as well to protect the apples from bruising. Apples stored in this way can be stored for months in such a manner. It may be that other fruits such as oranges, lemons, pears, coconuts, unripe bananas and tangerines could be stored in this way, experiments need to be taken to see if this would work. I recently put an apple in a draw in my office in about June and in September I opened the draw and ate the apple and it tasted fine, if a little wrinkled. This idea would not work for soft fruits such as peaches, strawberries, melons, plums and similar fruits which spoil quickly.

One of my son's friends from schools father, recently found a website on the internet called www.omlet.co.uk which sells chicken runs, chickens and everything you need to produce your own eggs. Very soon my son's friend came home from school one day and there it was, a lorry had arrived dropped off all the equipment and the chickens and everyone was tucking into fresh cheese and tomato omelettes, watch out for the cholesterol though, on the other hand if you are starving and have had no fresh food for weeks, forget the cholesterol and tuck into the eggs.

A couple of words of warning though, if you are thinking of tucking into a tender juicy chicken pie for supper beware, adults and children get very attached to their chickens, they are like friends of the family and I would find it very hard to wring my chickens neck, I might have to bring in one of my sister in laws to actually do the dirty deed, and of course not in front of the children please.

The other issue of course is foxes, they are very resourceful, in the end it wasn't my son's friends father that ate the chickens it was that sly old fox who had dug under the wire netting and climbed in and had eaten all the chickens. Apart from standing guard with a shotgun, the other method of making the chicken run fox proof is to ensure that chicken wire is actually all the way underneath the chicken run and is either twisted into the upper section or using a staple gun, staple the chicken run to the underside chicken wire. Believe me foxes are very adept at breaking into chicken coops, they have been doing it for hundreds of years. Please also note that German Shepherd Dogs can also be partial to a chicken dinner, there may well be bands of feral dogs roaming the streets who will be able to sniff out a healthy chicken at three blocks away, be on your guard.

For those of you with very big gardens, you could possibly keep, ducks and geese, which make good sentries and start cackling at the slightest strange sound, you could possibly keep sheep and goats and pigs, however your house will start looking like Felicity Kendal's house in "The Good Life," make sure you have several pairs of strong wellies. The problem with keeping all these animals is how are you going to feed them when the balloon goes up, unless you also build a grain silo on your land, and unless you are Prince Charles and perhaps own land the size of Cornwall it might be asking a lot of one's wife to start ripping up the rose beds to install a 40 foot grain silo and chicken runs and pig pens. We all did it in the last war though and we all, apart from those that actually fought in the war, were very healthy on the diet we lived on then.

Upon reflection maybe a small chicken run and a vegetable garden growing carrots, potatoes, beans, tomatoes and lettuces might be the maximum that most people with decent sized gardens could go to, even with the chicken run though you will need a hefty amount of chicken feed on tap to keep the hens laying for 3 – 6 months, unless chicken pie is on the menu, when the grain runs out. Again these are decisions and choices that individuals will have to make and all these choices involve a fair amount of time and dedication to see them through.

A. Canned Foods

Canned meats, various types, including Irish Stew, sausages, mince, pork, ham, baked beans, peas, carrots, fruit chunks, sweet corn, tuna chunks, red salmon, olives, sardines, potatoes, condensed milk, canned butter,(I think fortnum & Masons high class department store in Piccadilly in Central London, although very expensive to buy things here) beetroot, other vegetables, fish and meats (as available).

B. Dried Foods

Dried potato flakes / powder, all types of pasta, cous cous, dried fruits, figs, apricots, mixed fruit, sultanas, currents, sugar, powdered milk, dried meats, (e.g. Biltong and dried meats from South Africa and the USA), flour, semolina, rice, dried lentils and all other dried vegetables, cereals and fish, meats and fruits.

C. Foods Preserved in Glass Jars

Many foods these days can be found in jars such as preserved fruits in syrup, fish and meats in preservatives e.g. aspic and jellies. Some high quality butchers and deli's have these sorts of preserved foods, however they tend to be much more expensive than canned and powdered foods. Jams, marmalades and other preserved fruits.

D. Baked Foods

Such as Ryvita's, Carr's Water Biscuits, crisp breads, rice cakes, french toasts and many other baked bread and sealed cakes, biscuits, in tins and packets. These foods as long as they are kept dry should last for quite a long time although not as long as canned foods. My advice would be to eat the crisp breads, cakes and biscuits firstly, then perhaps the powdered dry foods and leave the cans and preserved foods to last as they will last the longest. Some of Captain Scott (of Antarctic fame) canned foods and whiskies from 1911 are still edible today 100 years later.

E. Bottled Drinks and Water

Some crates of still bottled water (although be careful of the weight on floorboards in your home as all the above food is quite heavy), it might be preferable to store all the water on the ground floor, or in a garden shed and put the lighter powdered and dried foods upstairs (if you have a house) and also to keep the canned foods (which are very heavy) on the ground floor. Bottled water has a maximum shelf life of about 6 – 9 months and should be kept in a cool and dark place if possible. Water will be one of the main essentials that no one can live without for very long. After an eruption, depending on the chemical makeup of the ash cloud, the tap water should be ok for a while. If the Authorities broadcast that due to fluorine or other forms of poisoning one cannot drink tap water then many serious problems could occur. I am sure that the Government has access to underground reservoirs and/or aquifers and wells which are underground water retaining rocks which act like a natural filter to water, which falls as rain and then percolates through the ground to these aquifers. The Government will then hopefully start water bowser deliveries and or set up stand pipes possibly linked to safe drinking water supplies.

In a real emergency canned and bottled beers can be drunk although not by children under six years old, you will also have bought some UHT treated long life milk which has a shelf life of about nine months to a year and which would be suitable for babies and young children. There are other forms of milk such as tinned condensed milk and powdered milk (although that obviously requires mixing with clean water before it can be drunk). I would advise buying a mixture of bottled waters, soft drinks, wines, beers, ciders, as much as you can safely afford and safely store as without liquids no one can survive for long.

Using rain butts to collect water will not be an option early on in the eruptive cycle as the water could be contaminated with sulpurdioxide or other poison chemicals and the rain would then be a dilute form of sulphuric acid, although it should be noted that this will be a very diluted form, it still, in my opinion initially would not be safe to drink. However after possibly a few weeks when the eruption has dwindled (hopefully) it may be possible to use your rain butt (connected to your high level rain water gutters) and then after perhaps boiling and checking the acidity, if you have a PH tester (this is an acidity tester, you can buy these, look on Google) it should be ok to drink the water. If the water is slightly acidic you could add some alkali tablets or water purifying tablets to the water to make it more potable. I will do further research on this issue and post my results on the website www.icelandicvolcanoes.co.uk as soon as I have any further news. I would also imagine that the Government scientists will also be regularly analyzing rainwater and reservoir water and that regular bulletins regarding the safety of drinking water will be posted on notices locally or loudspeaker vans will tour your area, or radio and TV will give their findings. If in doubt do not drink the water until a GP or an official from the Government or water board has confirmed that it safe to drink.

Lucozade, (for energy), Irnbru (for iron, essential for the body), Ribena, (vitamin C), orange squash (vitamin C), wines, beers, spirits, whisky, cans of soft drinks, bottles of cider, fruit squash's and all other sorts of bottled drinks which have a long sell by date on them.

There will probably be many other foods and drinks which I haven't thought of which can be purchased, however the main criteria here is the sell by date (which is basically only a guide) many products can still be eaten slightly after their sell by date but please beware in case the food is off and then could be dangerous, you do not want to get food poisoning if it will be hard to get to a hospital or doctors surgery.

I have recently been conducting some further investigations about the longevity of food supplies and I have come across an American website called www.survivalacres.com which is linked through another website discussing the various calamities that could destroy the world at any moment such as SARS, Bird Flu Pandemics and similar ideas, the idea being that many Mid West (mostly) US citizens are building bunkers and compounds so that when these 2012 style calamities occur they can hunker down for a few years with their huge supplies of long life foods and emerge and take over the world, or something along those lines and the proprietor of Survival Acres has a massive range of air and freeze dried foods, where they have managed to take the oxygen content of the foods down to very low levels so the food, if stored at about 70 deg F can last up to 40 years, the foods include all types of beans, mushrooms, fruits, eggs, flour, pancakes, meat stews, all sorts of wheat products and according to the accompanying information FDA tests have been done on these products and the vitamin contents have survived so that humans will be able to obtain their quota of vitamins and roughage and other dietary requirements from these foods.

I would have to point out here although I have had a good look at this website I haven't actually eaten any of their products yet, however in the not too distant future I will be purchasing some samples and will post the results on my website www.icelanicvolcanoes.co.uk I would have to say that the website looks very serious and I would imagine that the products do exist and according to the sales blurb the company has been in business for 15 years and is very popular with hikers and backpackers.

Most of the recipes involve boiling some water and reconstituting the food with hot water. There are probably other similar websites out there, I will try and source them and put them on the website. The only downside to Survival Acres seems to be that they do not ship directly abroad, I have a contact who is a shipping agent at JFK who could help with any shipping, and then there is the cost, these specially prepared foods are not cheap as they require specific production processes that most foods do not require. It is however an interesting idea as for lazy people who buy some food with a 40 year shelf life, you can just buy the food, put it in the larder and forget about it until the eruption starts, however I for one would not be content to do that. I would still use my normal system of purchasing canned, dried and pickled and baked foods and rotate them on a 6 month basis.

1. Rotation of foods & drinks

This will be a hit and miss affair as you will have to rotate the food every 6 to eight months to keep it fresh, basically you will have to ensure that you put stickers perhaps with a colour code on the foods, e.g. food bought now in May 2010 will need to be rotated (i.e. eaten) by about December 2010 so you will gather all the foods and drinks with the yellow tag (use a colour code for time periods) and put them back into your larder and food cupboards to be used by you and your family and then go out and replace all these foods and give them another colour code. Some of the foods will have to be rotated more quickly than other foods as cakes and some crisp bread will obviously not last as long as cans of foods. You will have to be very organised here as you will not want old foods which are past their sell by date left in your stockpile as they will be inedible when the volcano erupts. You will have to use common sense regarding all of the above criteria as if you live with a large family in a one bed roomed flat it would not be possible to store 30 crates of bottled water, rather you should store (in this situation) more dried and powdered foods and some cans as this will allow more sustainable food storage per square foot for this sort of property. I would envisage that there will be continuation of water supplies as the Government will either organise stand pipes in the streets or water bowser deliveries to notified sites during the period of crisis. In my opinion the provision of foods to the population will be the biggest problem that the authorities will face.

The question of how much food and drink to stash away is also quite important, If we go back to our average family of 4 persons eating 2 cans of food per day each, an average tray of cans from the supermarket contains about 36 cans therefore approximately the average family will consume about 2 boxes of cans per week which would equate to about 100 boxes / trays of cans per annum, the actual space required to store 100 trays is not that large and I would advise not putting all those trays of cans in one place in your flat or house as the extra weight could damage load bearing floorboards in some old conversions. The same way all the bottled water should be spread around the home, this practice would also help if a burglar broke in and stole some of your food, not having all the food and drink together might save some for you and your family.

I would also advise not leaving the food and drink close to sources of heat such as boilers, hot water tanks, radiators (unless they are turned off) or near a window where the sun's rays would be magnified by the glass and heat up your food and water. The ideal storage area might be a dry cool cellar or roof loft or similar space, or a dry cool outhouse, secure shed or even perhaps a garage or store room.

The dried foods, powders, flour etc should not take up much space in your home, however the biggest problem might be the storage of water. My hypotheses are that Central Government would have to prioritise potable (drinking water) as without drinking water everyone will die of thirst fairly quickly. I foresee that most reservoirs unless they are totally flooded with tephra, ash or collapse and break would still probably be useable as long as the filtration systems and fluoridisation and other health additives were continued to be added. Even if the main pumping system failed the Government should be able to organise stand pipes on most street corners fairly quickly, failing that regular delivery by mobile water bowsers would take place. Personally for drinking or cooking I would prefer to use a purer source of water hence I have put bottled water on my list of "must haves", however (especially in summer) the quality of bottled water will deteriorate after 3 – 4 months of storage. There is not much more one could do apart of to start thinking the way people thought during the Middle Ages and in Tudor Times when water sources were not clean and terrible diseases such as Cholera, TB and other diseases could be contracted from poor sanitation and poor quality water. What did they do, they fermented alcohol and drank mostly beer and spirits which would keep for ages and due to the fermentation process and the alcohol in the beer and spirits this killed all the bad bacterium and bugs in the water. I am not suggesting for a moment that you start giving babes and 7 year olds Holstein Lager for supper, however (as they did in those times) children were given watered down beer and wine (as they still do today in France and much of Southern Europe) as it kept them safe from waterborne bugs. Alternatively (as long as the bottled water was not too old) give the children and babies the bottled water after perhaps boiling it firstly and letting it cool down and the adults perhaps drink watered down beer, cider, wine etc to keep the supplies going longer. There are so many imponderables to consider here that this part of the book could go on for chapter after chapter, however in essence as with many of the points that I am raising in this book adults have to make adult decisions for themselves and their own families and friends, there will be minimal input from the "nanny state", however I am sure numerous medical bulletins from the Chief Medical Officer of the NHS will be broadcast on a regular basis to help and advise individuals and families about what is healthy, advisable or dangerous to do. I am not trying to supplant or over ride Government advice, however this book will attempt to help people from being caught unawares and give them a head start in preparing for all eventual possibilities.

A lot will also depend on the severity of the volcanic eruption and length of time the ash cloud is over the

UK and EU, also a lot depends on the chemical makeup of the ash cloud, is it a poisonous Laki gas cloud or is it from a Yellowstone style super volcano that pushes the Earth into a 2 year nuclear style winter, there are many gray areas, like a small EJ style eruption and situations in between that I possibly could not cover all eventualities. I hope to have my website "www. icelandicvolcanoes.co.uk" up and running by the December 2010 (this website is now mostly active and up and running) which will have an updated section which will be password accessible and will give the latest advice that I can get from the volcanologists, scientists and other professionals on how to use your stored resources to your best advantage, whether it is safe to venture outside, what vegetables one could grow in your garden or where Government controlled shops and depots can be found and the times that they will be open. Whatever else happens you can be assured that I will keep the website updated as quickly as possible as the ash cloud changes. If the power supplies are sporadic it might take me longer, however I will do my best to keep everyone updated with the latest news.

Returning for a moment to the drinking situation, I do feel that with stand pipes, water bowsers and bottled water things shouldn't be too bad, canned beers, ciders etc, should last for about 6 months or so. The other alternative, probably only suitable for those who have their own house is to build a "break tank" at home. "Break Tanks" are tanks made normally of strong durable hard plastics which the cold water mains feed into (mostly in modern and high rise blocks of flats), if the mains fail due to ash cloud activity or super volcanic activity, then a full break tank could hold hundreds of gallons of water, however stagnant water would not remain healthy for probably more than a month or two at the most, one could add water sterilisation pills to the water, however it could be useful if several family members go to the stand pipes using buckets etc and refill the break tanks on a regular basis. All of the above ideas could be used in conjunction with each other in a worst case scenario, firstly use the water in the break tank, secondly use the bottled water (saving some for babes and small children) then falling back on watered down beers, ciders and wine. Once all has gone other water sources could be used (with care) by first putting in water sterilisation pills to clean the water (as they do today in many parts of Africa where there are contaminated water sources which are the only source of drinking water).

2. Medical & other necessary supplies required

Apart from food & drink there will be many other necessary items that one would need to store away for the time when the power fails and no one comes to help, no police, doctors, ambulance, plumbers, electricians only you and your family fending for themselves in a frightening dark and dangerous world.

Although it cannot be worked out exactly what level of support if any comes from the local authority or Central Government after the eruption, this all depends on what level of magnitude the eruption becomes. In the worst case "Yellowstone Eruption" the emergency services would quickly be inundated with cases and be so overstretched that they would collapse, even if an ambulance could get to you, all the hospitals would be overflowing with casualties and you would be left on a trolley in a corridor for hours, if not days until an overworked exhausted surgeon got to you, just look at what happened in Haiti after the earthquake, everything was chaotic all buildings flattened and those that were not flattened were overcrowded, no services, no water, electricity or any basic help, it was only when many nations flew in field hospitals and teams of surgeons and medical equipment and generators that anything started to happen.

It might be sensible to find out where your local GP's live as if the worst happened and no one responded to 999 calls (that's if the phone lines are still working) you could go round to their house and ask them to come and see the injured or ill person.

I would recommend that the following be purchased for the average family (of 4).

1. First Aid Kit

To include bandages, medical scissors, sutures, gauze, plasters, splints, medical tape and all the normal medical extras, e.g. cotton wool, corn and blister pads etc, etc.

2. Pills & Medications

i. Headache & migraine pills, e.g. Neurofen, Ibuprofen, Sulpodene, Panadol, Aspirin and

Paracetamol. These products normally last for about a year, so don't forget to colour code them for recycling after, say 6 months to your own medicine cabinet for you and your family to use and then replenish your emergency cache stocks. I think a 6 month cycle for medicines generally should be ok, however please do read the expiry label on the bottle or packet or tube of medicine and ensure that half way through the expiry time that you recycle the medicines and use them and replace your stocks with fresh medicines. This is a very important point which I cannot emphasize enough is highly important. You do not want to have survived the first 3 months of a volcanic winter only to have family members come down ill or incapacitated somehow and then find that all the medicines are out of date and cannot be used.

Also, please ensure that you do buy enough of these medicines, for a family of four for a 6 months to a year you would need to budget for a packet of pills every 2 months as a minimum, therefore for example, for a family of four you would need 6 large boxes (of 24) Sulpodene tablets to last the year.

ii. Hay fever, sinus and asthma remedies, such as anti histamine's (puffers), Piriton and all proprietary brands, such as Sinutab and Flixinaze (which is a mild steroid spray for sinus's)

iii. Women's Preparations, such as vagisil, preparations for thrush and gynaecological problems, (please ask your GP for fuller details than I can give)

iv. Thyroid Preparations, such as Thyroxin.(seek GP advice)

v.   Muscle Problems : Deep Heat, muscle relaxant sprays and lotions.

vi. Osteoporosis: seek GP advice, for various remedies on offer.

vii. Gastro Problems: such as anti acids like Milk of Magnesia, Gavascon (for heartburn) and indigestion, Sena Pods (for constipation) and preparations for diaohrrea and other stomach related problems (ask your pharmacist for advice here)

viii. Heart Problems: This is going to be difficult unless you have a very good relationship with your GP and he / she will allow you to take on prescription enough heart pills for a year e.g beta blockers, stattins and other blood pressure pills that cannot be bought over the counter. If you explain to your GP that you might be going on a 6 month holiday around the globe then maybe you can convince them to give you a large enough prescription to survive 6 to 9 months. Hopefully there will still be some vestiges of A & E departments working, however the problem might not be the A & E staff, the problem will be getting the pills from a factory which has closed down due to the ash cloud activity and the staff are at home (like we all will be) and as they say "no one will be manning the pumps". I hope I am wrong and Governments decide that strategic industries like pharmaceuticals and other factories making medicines and related products are given special help or emergency procedures to keep them churning out medicines and pills in the midst of a global emergency.

I would advise to be on the safe side stockpile as many urgent pills and medicines as you can reasonably afford and safely store at your home.

ix. HIV Medicines. The same goes as number viii for these products.

x. Hepatitis B and C and diabetes. The same goes as number viii.

xi. Malaria, typhoid, yellow fever, dengue fever, west nile fever and all other noteifiable and serious illnesses and diseases. Please see my answer to no. viii.

Children's medicines: Calpol and Medinol, Trixilix and all other children's medicines, purchase as much as you can sensibly afford and safely store (out of reach of children obviously, as with all the other medicines described above)

xii. Skin Creams and body lotions

It will be difficult going out to Harvey Nicks and Harrods to buy the latest Clarin's preparations if a 2 metre ash layer is on the ground and red hot spheroids are raining down on the No. 13 bus. My advice, without re mortgaging your home is again to purchase enough of these skin creams, lotions and potions to keep you going for about a year and rotate them every 6 months.

Definitely buy creams like E45, Vaseline, hand creams, sun protection cream, for when the sun finally peeps out from behind the ash cloud in a year's time you won't be able to pop into Boots to buy your factor 30 as Boots will probably have collapsed under the weight of tephra or been looted by a mob of discontented people who did not read my book or could not afford to buy the creams in the first place.

xiii. Soaps, shower gel, shampoo: This will all still be necessary, if some water is available you will want to wash, shower or bath occasionally during the year, although probably as we will all smell pretty atrocious (as in the eighteenth century when ladies and gentlemen of fashion had pomades filled with sweet smelling herbs to protect their noses from peoples awful smells) so anyone who is too clean will be looked at with suspicion of having a cache of VO5 or Dove shower gel.

xiv. Toothpaste, dental floss, denture fixatives, tooth picks: We will all have to try and protect our teeth, especially the teeth of children, as fresh fruits and salads will be in very short supply, then maybe extra vitamin C & D tablets will have to be given to young children and teenagers to keep their teeth healthy and from protection against diseases like rickets and scurvy which follow a lack of fresh food and one of the symptoms is teeth falling out. That is why earlier in the book I recommended an apple barrel as per ships of the line in the 18th and 19th century, that's why us Brits were called Limey's because under various admiralty regulations a barrel of limes to be sucked and eaten by the crew to stop rickets and scurvy went aboard all Royal Navy vessels, possibly a barrel of lime juice could also be used to combat bone problems.

3. General household items

You will not be able to pop down to the shops to purchase any household items as firstly most of the staff would be cowering at home glued to CNN, BBC or SKY and / or the shops will quickly be cleaned out by everyone who has not planned ahead, panic buying everything in sight. The problem might well be that key members of the distribution services, i.e. HGV Lorry drivers, ferry staff, rail workers operating signals and other key services may not turn up for work or be able to get to work, therefore even if containers of food and household items did get delivered it would be a haphazard and not a regular supply, much like Austerity Britain in the 1940's when if people saw a queue they joined it for whatever it might be, a shipment of shoe polish or New Zealand Lamb. This was called "the rationing mentality," as if goods and services are restricted the demand for them increases therefore you will always have a huge unrequited demand for most items until supply equals demand, (that was a quick lesson in Keynesian Economic Theory).

The items that you will need to purchase to keep you going for about a year will obviously vary (as mentioned before) will depend if you live on your own in a 13th floor high rise or you are a family of 6 living in a huge detached house in Chalfont St Giles, you will have to use your own common sense and try and build up your stockpile of food, water, medicines, campingaz canisters and heaters, household items and anything that you think is an indispensible part of living. We have become so interdependent on popping out to the 24 hours supermarket or convenience corner store that we have forgotten what it is like to really try and become self sufficient, even in the comedy drama "The Good Life", the self sufficient couple often seemed to be having TV dinners with Gerry and Margo next door in the programme.

What we need to do is sit quietly at our table and write down all the items you would consider taking to a Robinson Crusoe Island (for you younger readers think about being marooned on "Lost"), and no you can't take a time machine with you!!. List out all the items that you think you would need to survive for a year without any outside help at all, then thinking about your budget cross check what you can easily afford, then think about where you are living, a studio flat in Hammersmith or a 4 bed roomed semi in Croydon and try and work out a practical list that is both affordable and available and won't break your floor boards.

Then decide over how long you are going to take to build up your supply of items needed and whom you are going to tell about your supply cache. I feel if all this is done slowly, carefully and methodically then you have a chance of surviving the next Icelandic volcanic eruption.

Section 2

How to survive the upcoming Icelandic strato volcanic eruption

By Mark Reed

Copyright Mark Reed 2010

2. Household items needed

ii. Lighting

If matters get serious and power cuts start becoming frequent then the way of life as we know it now will become impossible to maintain. Without regular mains power, items like computers, TV's, CD's, MP3 players, microwave's electric cookers, gas fired central heating (which rely on an electric controller to function) will all cease to operate. I am hoping the interruption to our lifestyle will only be between 3 - 9 months depending on the severity of the eruption and where in the world it was. The events of 535 – 536 AD were severe across the globe, however they didn't have oil and gas powered power stations nor the internal combustion engine, nor did they have any real way to store food, as we do with cans and freeze drying so the effects on the Roman Empire would have been much harsher and quicker acting than the effects will have on us in today's modern world.

However let's start by assuming the worst for a while, without mains power everything in our world comes to a standstill, where I currently work in South London when we have a temporary disruption to our computer server, which is fairly often, I see work colleagues just sitting there staring into space awaiting the IT guru to sort out their computer glitches, their world is so inter dependent on computer technology that people seemed to have forgotten an age where ordinary people made do (all of their lives) without computers, the Internet, tweeting, texting, e – mailing, and blogging. When I now go on the Underground to work, as I do every day, I notice more and more commuters instead of reading a good book (like I do) or the snooze paper, they are playing computer games on mini keyboards or writing long messages on BlackBerry's to be sent once the train gets out of the tunnel. Now I hear that TFL are thinking of using the Underground to mount a Wi-Fi network, so soon there will not be any sanctuary from tweeting, texting, yelling commuters and their ever increasingly complex, games, gadgets, phones and toys. Let's pretend (perhaps in my case not too unhappily) that this all ends, what do we do to survive, how do we cope?

Torches, candles and other lights.

These days you can buy wind up torches and radio's which seem to last forever on a couple of turns of the handle, a fantastic invention as if the worst comes to pass there will be no corner shop with a packet of AA Duracell Batteries to buy, firstly they will probably sell out pretty quickly, secondly and in all probability the shop would get ransacked and everyone would make off with their handful of batteries and other goodies. Obviously it wouldn't hurt, as well as purchasing a couple of wind up torches to also buy a couple of battery powered torches, in case you lose the wind up torches. Campingaz canisters can also be made into lanterns to hang up which give good light, however they would quickly be used up and they should be conserved for cooking. Paraffin (in days gone by) was used extensively for heating and for lighting, such as miners lamps and similar items, I am sure with a little hunting around car boot fairs or camping stores that some old miners lamps or other sorts of oil lamps can be sourced and the correct sort of oil could also be found if one looks around. The large scented candles popular with single ladies eating Cadbury's Flake chocolate bars and romantic couples having dinner by candlelight could be used as the large ones take about 4 days to burn. Some of these large candles are not too expensive and I would strongly recommend buying some. Shops like IKEA have lanterns in which you can put the candles as especially with kids running around you would not want to leave a naked candle burning or a fire could break out and I doubt if any fire engines will be on call especially if the phones are dead too.

If there are prolonged power cuts I can foresee real problems here as we are not really used to living without gas central heating, electric lights in every room, TV's, radios, MP3 players and all the usual comforts of home, in the streets outside where you live are nice bright street lamps, what happens in your street at night if all these street lamps suddenly are not on at all at nights (I understand that this might also happen as the Coalition Government and local authorities are trying to save money on street lighting in the UK), the world just became a much more frightening and dangerous place. There are currently parts of London that I wouldn't venture into late at night with street lighting on, what it will be like without street lights, heaven knows. I would imagine that unless a group of people are going in the car (if you have any petrol that is) and you might venture to a friend's house or to the community centre, where hopefully the Government will have a few arc lights up to throw some light on an area where they are looking after people.

For the rest of us, we will probably be huddled around some sort of live fire, don't board up those beautiful old Victorian and Edwardian Fire Places any more, in fact if you have one get a chimney sweep in to clean the flue or get the flue repaired, unboard it and try it out. A wood or coal fire can give out light as well as warmth and there is something nostalgic and cosy about huddling up before a real fire, expect a major baby boom nine months after the blackouts, (it happened last time in the early 1970's), what life will be like without "X Factor" and "Britain's Got Talent", I really couldn't imagine, maybe people will learn to read again, and no I don't mean OK Magazine or Chit Chat, I mean Dickens, Dostoyevsky, Voltaire, Nietzsche, Chekov, Stefan Zweig, Agatha Christie, Jeffrey Archer (well maybe not all that list has merit). We will all have, for a while a lot more time to reflect upon our lives, what are we doing with our lives, where are we going, are we achieving anything purposeful in our daily lives or are we all simply machines plodding backwards and forwards on the Underground every day. This spell without TV, without daily work might actually brighten up many of our lives and give us something to strive for, for a while, whilst the emergency continues, we will have to fend for ourselves, look after our families, interact with neighbours, friends and form committees for self defence, getting food and water and other hunter gatherer sorts of lifestyle. I don't think it will be exactly like a Jane Austin novel, such as Pride & Prejudice, however it might give some of us a short time in which to carry out introspective reflections on our lives, in some macabre way I for one will find it quite an interesting challenge having to cope with real life, all the real dangers and trials and tribulations of hunting for wood, food, water and the basic necessities of life which for a few months (hopefully not too long) we will all have to live on our own wits.

We will however have one big advantage over most of the population in general, as we would have read my book and at least at the end of a hard day collecting logs on Hampstead Heath we will return home to a sumptuous meal of fried spam fritters and Walkers Crisps, followed by some well aged Ginger Beer, or if we are really lucky some cool Gaymers Pear Cider, sounds good to me.

Getting back to reality for a moment, if you do have a proper useable fire place with a working flue, you should be ordering, coal or coke now and start collecting a big pile of cut up logs and put them around the back of your house or maisonette out of sight from passersby. You can still buy oil lamps if you look on the internet, there are all sorts of Xmas metal lanterns for sale at Xmas fairs across the country and various oil and paraffin lamps and candles should be collected as 3 – 6 months is a long time to go without light if the worst comes to pass. I am hoping that some sort of power will be on possibly for a few hours a day as they might have to have rotational power cuts across the country, however if that does occur and you have rechargeable batteries or even a car battery linked to a battery charger there may be the possibility of having some sort of lighting system in place. Again you will have to work out what is feasible and what is safe and what is best for your own situation, we don't want to see a roaring log fire in a top floor high rise block of flats with no ventilation or people will be dying of carbon monoxide poisoning and fires will be breaking out across the country. Please ensure if you do have a wood, coke, coal or peat fire ensure that windows are open for ventilation and that your chimney and flue are working properly and that before you all go to bed that the fire is put out and not left unattended, it only needs a couple of sparks from a spitting green log fire to set fire to carpets or curtains and then your whole house or flat will be up in smoke as there probably will be no fire brigade to call, if the phones are working that is.

Cooking

While the gas is still flowing and the electricity is still on you will have no problems cooking your canned, dried, pre packaged and frozen foods, I would definitely recommend eating all the frozen food and cold meats and other fresh foods that you have in the fridge firstly as when the power goes off and the gas is cut off it will all go rotten pretty quickly. What do you do when all power and gas is off.

My first choice would be to use campingaz canisters which you can buy from such shops as Blacks Leisure, Millets Stores or other reputable camping and outdoor shops. These canisters come in various sizes. The larger canisters can be attached to a sort of portable gas cooker grid and can be used indoors as long as you properly ventilate the rooms that you are cooking in. For a family of four cooking 2 meals a day I would imagine that the large canister if used cautiously and not boiling cups of tea every 5 minutes, could last for about a week to 2 weeks. Therefore in a perfect world you would buy 52 of them and you would have a year's supply of cooking gas.

I would have to state very clearly here that the safe storage of these gas canisters (Campingaz is a registered trade mark of canisters of gas used mostly for outdoor cooking and lighting) is very important, these gas canisters could, if improperly treated or exposed to direct heat explode and burn down your house or block of flats and injure adults and children alike. The canisters must be stored in accordance to the manufacturer's instructions, which probably state that they must be kept out of direct sunlight and excessively high temperatures and also in a well ventilated room in case they leak gas into the atmosphere. If you have a lockable safe outside shed or garage that might be the safest place for the canisters, or perhaps a dry cool ventilated basement or similar area. If you live in a block of flats please be extra careful as a gas escape or explosion in a confined space could well set fire to the whole block. Again if you have a communal shed or garage block maybe it would be best to store the canisters outside.

These canisters can also be used as reading lights by attaching a special lamp fitment, again please be extra vigilant with small children around as if these lamps turn over they could be very dangerous.

With your campingaz cooker you can boil, fry, poach in fact apart from baking, roasting or toasting you can cook almost anything from fresh fish on a griddle to baked beans. Boiling water to purify it can also be done and I am sure with your extensive combination of tinned, freeze dried and powdered foods you will soon be cooking up cordon bleu cooked food for all the family, Gordon Ramsey and Jamie Oliver eat your heart out, in fact

I might write to both of them asking for separate recipes for this sort of cooking, if I get a positive response I will post it on the website www.icelandicvolcanoes.co.uk, that's if there will be any power to download the recipes.

Other forms of cooking would be gas operated barbeques, again if you are bringing these into the house or flat make sure they are properly ventilated, we don't want to hear about cases of asphyxiation in the papers. The large gas canisters that come with most barbeques can be bought from most petrol stations and should last 2 - 8 weeks for a family of 4 so you will only need 26 of this size gas canisters. There are much smaller campingaz canisters that can also be bought, if you live on your own in a studio flat you might only want to buy the smallest gas canister, so buy about 40 of them to keep yourself going for about 1 year. If you live in a house near woods and you have an outdoor barbeque with a grate you can collect firewood and cook your own meals over a flaming grate.

If you are fortunate enough to own a wood burning stove like an AGA or similar design and you live near a wooded area then you will be able to carry on as per normal and you will not see much disruption to your daily life.

Whatever form of cooking you decide or is appropriate to your situation you will have to be extremely careful regarding the safety implications and if possible obtain a powder or gas style fire extinguisher and a fire blanket, to throw over the fire, to have near your cooker to extinguish any fires that might occur. Be very careful once you start to cook that you only take the correct amount of food from your stock pile of food. You do not want to cook great feasts for the whole street night after night only to run out of food after a few weeks and then have to go and queue with everyone else for 3 hours to get a stale loaf of bread and a small bottle of water to subsist on. Try and define how many cans or packets you are going to use for each meal and ensure that you stick to your plan, that's if you don't want to run out of food half way through the emergency.

I definitely believe that after only a few hours after the emergency has been declared that all stocks of campingaz and other gas cylinders will have disappeared from the shelves and would urge you now to go out and buy some campingaz or barbeque, large gas cylinders from the petrol station. If you are cooking using wooden logs or coal or coke fires please make sure that someone is watching the fire after everyone has eaten, a spark from a log can easily set fire to curtains, rugs and other soft furnishings, fires must always be watched carefully to avoid fires.

On the positive side, with regards to early nights with no TV, as with the power cuts in the 1970's it might encourage a baby boom nine months later, after all couple's will have to do something to while away the long winter nights with no TV or PS3 to play with. Let's hope the maternity wards are up and running 9 months after the eruption.

I would also invest in a couple of car batteries, if you have the space they might be useful as a back up power supply or for other emergency electrical power if required.

General household items required

The list is practically endless and many householders and flat owners will already have many of the items that I am now mentioning here, obviously if you already have a serviceable item at home, as long as it's not on its last legs and will keep going for 6 months to a year do not rush out and buy a new one, that would be a complete waste of money (and your valuable time).

Tools for home and car

Screwdrivers, both straight head and Phillips cross head, (2 of each if possible), hammer, nails, screws, raw plugs, hand driven mechanical drill, cordless power drill (with 3 fully charged battery packs for backup), sand paper, panes of glass, duck tape (about 6 rolls, very useful), spanners, light bulbs (enough to replace every light in the house 3 times over), RCD's (residual current detectors) to be plugged into your fuse board (old style fuse boards) perhaps a socket set, string (about 4 balls), clothes pegs, clothes line, wire, fuse wire, adjustable spanners (monkey wrench) electric testing screwdriver (one that lights up when you touch a live wire), battery operated smoke detectors (and a plentiful supply of spare batteries) carbon monoxide fume detectors, water fire extinguishers, foam fire extinguishers, powder fire extinguishers (you will have to go to the various fire brigade websites to note down which extinguishers to use with which fires, e.g you do not use a water extinguisher for a chip pan fire, you need a powder or foam type extinguisher, also for electrical fires you do not use water only dry powder types) electrical tape, electric wire connectors (plastic item which connects individual electric wires). Toilet Plunger, toilet brush, pincers, workbench (like a Black & Decker workbench), files, rasps, wood plane, wood saw, (2, one small, one larger), cold chisel, vice, clamps, clear mastic, (3 tubes) for waterproofing bath edges, showers and kitchen areas. Maybe even a small bag of cement, plaster and sand and a couple of panels of plaster board and a few roof tiles would be useful as well as plastic or canvas tarpaulins to cover holes in roofs and walls. Also clear plastic sheeting, (in case windows get smashed so you can stop draughts and rain coming in and still see out of the windows. Also some plywood or marine ply to board up damaged areas of your property and make them secure should it become necessary. You will have to use common sense with the above list, if you live on the 20th floor in a high rise you probably won't need most of the above, you will probably be too tired after walking up and down 20 flights of stairs 3 times a day (in the case of power cuts) to do much handy man work.

On the other hand if you live in a crofter's cottage on the banks of Loch Lomond in Scotland you could probably do with most of what I have described above as if anything untoward does happen you will only have what is with you to sort out any problems which might arise.

Car tool box with basic tools to do minor repairs on your car, one of my previous jobs was a mechanic for Peugeot and Lada so I know a little about car maintenance, and no, please don't all drive over to my house to ask me to repair your car,(no garages will be open), things like spark plugs, spark plug spanner, brake fluid, engine oil, automatic transmission fluid (automatics) gearbox oil (manual gear boxes) oil filter cartridge, distributor, workshop manual, wheel brace (for removing wheel nuts to change a tyre), 2 spare tyres, car jack, axle stands, car inspection lamp and long lead, emery cloth, spare headlight, brake light and indicator light bulbs, timing chains, fan belts, brake pads, water pump, thermostat, rubber timing belts, fuses, alternator, coil, condenser, emergency warning triangle, rubber latex gloves (one box) emergency windscreen (plastic clear view type), GPS Sat Nav that plugs into car lighter socket. Spare car keys and fobs, grease, WD 40, 3 in one oil, grease gun, spare car battery all other car parts that your local dealer thinks (and you can easily afford) to keep your car going for several months without access to a regular garage for servicing and repairs. Just think, the Government will have to cancel the annual MOT tests for a while, as there will be no one there to monitor the annual tests, that will save everyone £ 54.00 each and possibly tax discs too, check with the DVLA website (if its working) first though, I am not asking anyone to break the law.

Some petrol might be delivered sporadically to UK petrol stations and you might want to have the car on standby with a full tank of petrol in case of a medical emergency or other emergency to transport people or pick up supplies or to visit friends and family elsewhere in the country. This is going to be a tricky one to monitor as you will probably feel safer using the car to get about, however after you use up your full tank of fuel, that will be it, unless you can safely store some jerry cans with petrol you will not have any means to get anyone to the doctors or to a hospital in an emergency. Be very careful about how and when to use your car.

Other household items

Plenty of blankets, (for when the gas central heating goes out in winter and the house is freezing cold) enough for all the family and especially enough for the elderly, infirm and babies and young children. Sheets, pillows, pillow cases, duvets, maybe a couple of spare single mattresses (in case family or friends visit). Toilet Rolls, (about 10 crates if you have the room) when they run out you will have to use old newspapers, so start hoarding all old newspapers, we always seem to have a huge pile of old newspapers in the house for some reason, now I know why, my wife was always expecting an Icelandic volcano to erupt.

Toothpaste, all household cleaners, bleaches, washing up liquids, dishwashing Salts and other items (the power might stay on for a while before blackouts start becoming regular items). Washing Powders and liquids, soaps (for washing people), shower gels, shampoo, scissors, nail cutters, cotton wool pads, and all the usual ointments, medications and other items that normally clog up the bathroom and medicine cabinet.

Brooms, mops, buckets (about 5 these will be used for a variety of things), dustpan and brush, spare Hoover Bags (unless you have a Dyson Hoover), black bags (about 20 – 30 rolls of 50 to keep you going for 6 months to a year), cling film, and all other normal household and kitchen items. Can Opener, bottle opener, corkscrew, matches (30 boxes of large matches). Candles (about 20 boxes) and or some of those huge candles that burn for about a week (which I mentioned earlier).

As stated earlier there is no set list of items that you must have, however with a family of four to feed and you have forgotten or can't find the can opener, a small problem in today's world might (in the worst case scenario) be a matter of life in death in the Post Volcanic Ash Cloud World. I would urge all householders to designate one person to make a list of all items that are required, food, medicines, household items, car spares and then let another family member vet this list, and then try over, say a three month period go out and purchase all the items on the list. Then (as mentioned before) set up a six month rotation plan to keep the food fresh. I definitely cannot think of every item needed, you will have to take responsibility for protecting your own family against the possibly much harsher environment and world that will emerge post apocalyptic (or post volcanic) Europe. It may be that in the actual event of an eruption, that the effects could be sporadic, huge ash clouds one month and a quiet period for another month and then further eruptions, so that you might start using your supplies and then find that everything calms down and supplies start getting to the shops and things get partly back to normal only for the whole sequence to start again. You will have to be pragmatic, try and restock your supplies when the shops re open, perhaps not everything on day one, spread your purchases over a three week period maybe.

CHAPTER 3

Surviving the Initial Effects
of the Strato Volcano

So now you have purchased all your food, household items, medicines, car spares, you have divided the food up and made several different caches of food (in case burglars break in one day and steal one of your caches). Maybe you have hidden some of the food and materials in a separate garage you have rented or somewhere safe away from your house, these things will be up to you to decide how to plan for the worst case situation.

Hopefully it will be a couple of years before anything major starts, there are so many imponderables that it would strain the brain of a super computer to work out all possible eventualities. Let's take a scenario (that has happened before over Europe in the last 230 years), this is not a super volcanic eruption, this is a "Laki" eruption (Laki being a huge volcano that sits under a glacier in Iceland) that last erupted in 1783 – 1784 and which estimates say killed 6 million people worldwide (as discussed earlier in this book). Lets imagine that Laki erupts again in November 2011 and emits huge amounts of sulphur dioxide and hydro fluro carbons and that the news forecast state "a huge poisonous fog cloud is heading towards, Norway, the UK and Europe, we (at the BBC) advise everyone to stay in their houses and put wet towels across any draughty openings in doors and windows, if you have to go out please wear suitable breathing apparatus eg a special nuclear / biological army style gas mask and fully cover your body".

What would the effects of this broadcast have on the UK population, if say the fog was going to hit the UK in 72 hours time, I predict there would be pandemonium, panic buying, looting, petrol shortages, huge traffic jams everywhere as people tried to get back to their homes, family homes or tried to get to elderly or infirm relatives to help them or take them back to their houses. The Government apart from putting a few troops with riot gear on the streets and to protect public buildings like the Houses of Parliament and Buckingham Palace and the various Ministries and 10 Downing Street to stop looting, would be pretty powerless to do much initially. I am sure their first words of advice (don't forget this is November and its starting to get very cold at nights) would be "stay calm, organise bottled water, food, warm clothes and blankets and be prepared to stay indoors for about 2 weeks", however if everyone (there are 62 million of us) starts rushing off to ASDA, Waitrose, Morrison's, Tesco's and Lidl, after about 12 hours there will be nothing left in the shops to buy.

So let's say the majority of the population manage to get a few trays of canned food and a couple of crates of bottled water and a few blankets, then the fog hits and the roads are empty, the towns, cities, schools and motorways are all empty as everyone waits inside their houses praying for the evil smelling yellow poisonous "Laki Haze" to blow away. However as in 1783 it hangs around as there happens to be some high pressure over the UK and Europe so there are no strong winds to blow the sulphurous haze away, what then?

Whomever was on duty at the power stations would need relieving after 12 hour shifts, this would involve army units carrying biohazard suits getting to all the UK power stations and taking the relief workers in, suited and booted, and collecting the outgoing team and delivering them all home. Let's say, optimistically that the army can get to 50 % of the power stations, however we will all be at home with our central heating on, TV's, radios, laptops on baths running, microwaving, cooking, etc, etc and let's say that 10 % of the power stations go offline due to lack of staff to run them, the National Grid would not be able to carry the amount of amps and volts needed so mains fuses would trip, industrial sized capacitors in sub stations would overheat and huge blackouts over much of the country could follow. Let's say that the Government then ordered more army units in biohaz suits to try and find the National Grid engineers (they could be anywhere) then they will have to get them to sub stations all around the country to repair the tripped fuses and capacitors and other electrical repairs, don't forget all this is happening in the midst of a huge yellowish pea souper of a poisonous fog with temperatures plummeting and snow falling thick and fast, I am not even considering whether or not the spare parts needed for the repairs would be sourced, just look at what happens today when a street of houses loses power due to a short in the street, it can take a few days sometimes to get everything back to normal and that's with everything and everyone in place and doing their job correctly.

Let's also throw into this cheery concoction the fact that gas supplies might be intermittent due to problems with pipe breakdowns in Central and Eastern Europe and no one around to fix the broken pipes as no one has any biohazard suits or breathing apparatus. The UK only has a week's strategic gas supply and about a week's strategic oil supply (Germany has about 6 weeks supply of both). Therefore let's assume that key people are not in place and huge rolling power cuts and gas cuts for 7 or 8 hours a day become commonplace what happens to you and me.

We would have to quickly eat all the food in the freezer before it became rotten, so for a couple of days (subject to some gas and electricity getting through we would all eat well). Then those who didn't buy my book will be eating their cold cans of baked beans with sips of water in the dark, whilst those who bought the book would be feasting on hot tomato soup followed by hot Fray Bentos slices of meat with crackers and maybe an apple or two for desert, with maybe their wind up radio broadcasting the sad news that several army patrols had gone missing as they didn't have sat navs and were looking for grid engineers to try and get power restored in the West Country and a container ship collided with a ferry in the English Channel due to poor visibility, and other such cheerful news stories.

After about a week people who hadn't prepared would have to take huge risks and go foraging for food in the fog, they would probably put scarves over their noses and mouths so that they could breathe in the fog, although it is doubtful whether a scarf would protect you from the poisonous gases for very long. There would be more looting and the army and police would be patrolling wearing protective clothing and gas masks. After about a week or two at the latest one would hope that the Government in conjunction with Operation Cobra (response to a terror attack or major emergency situation in the UK) would start distributing basic foods and water to people street by street, though if they have enough bio suits, enough lorries and supplies is another matter and if they can distribute to outer London and provincial areas is still another matter.

Some people will start to drive around trying to find shops or supermarkets or Government distribution centres where some basic handouts are taking place, however I would imagine as peoples hunger and thirst took hold these distribution centres would quickly be deluged by thousands of people wanting water and food. The biohaz suits that were made to protect police and troops for a gas or nuclear attack are probably 25 – 30 years old now and I don't know how many of these old suits are left in MOD warehouses, or if the MOD actually know where they are, just a small point.

Lets now move forward, lets propose that the majority of the population would have survived the "Laki Haze" either holed up in their flats or houses or the Government might have opened schools and community centres (like the Americans did when Hurricane Katarina struck New Orleans in 2005) and that many thousands of people were in these community centres and schools being watered and fed by the Government, who had managed to scrounge and borrow supplies from manufacturers whose factories were lying idle but there were some stocks of food and drink waiting for distribution so the Government probably got very lucky here. Maybe some relief ships from Asia or South America got through and somehow got unloaded at the container ports, however we would need to stretch our luck a very long way here to believe that we would get such a lucky break.

I feel that some people like the homeless, people caught in the open, on camping holidays, some farm workers would have succumbed to the poisonous fogs, however I don't think there would have been a large amount of casualties initially comparable to the Laki Haze victims of 1783 – 1784. Don't forget that the population of the UK in 1783 was probably half (i.e. 25,000,000) as opposed to 62,000,000 currently. However after a week or two as people had to go out to try and forage for food and water across the country that is when most of the casualties will occur as people will only have scarves for protection and the fog will penetrate the scarves and many people will then succumb to the poison fog or get such bad respiratory illness that they choke up the hospitals with thousands of casualties coming into A & E every day. Hopefully the Government will have set up some distribution and safety areas by then so it may only be a case of getting to the local community centre before people find help.

After the Haze has dissipated due to the winds changing direction or a lessening of the eruption, other factors will then come into play. I doubt the UK will be showered with red hot molten spheroids as the Icelandic volcanoes are not exactly super volcanoes like Yellowstone and Tambora, they are smaller in size and therefore have a lot less magma to eject into the atmosphere. However as Laki is one of the bigger Icelandic volcanoes, I predict the amount of ejected sulphur dioxide gases will have a localised cooling effect on the Northern Hemisphere of about 9 months to 1 year (maybe 2 years at maximum). Therefore if the eruptions start in October then by November the winter might start to really bite with thick snow and ice and plummeting temperatures. We could have a winter comparable to Continental Canada or Russia with lows of – 30 deg C, which would mean that the Thames would freeze over and many water mains would burst. Pavements and roads would possibly be sheets of ice and snow, I doubt (due to the earlier Laki Haze) that the local authorities would be out gritting and clearing roads and pavements as the staff would not be in place.

We have to remember that we live in the UK where all the trains stop running when "the wrong sort of snow" or "wrong sort of leaves on the track" brings the country to a halt. Recently we had a very cold snap between Dec 2009 – March 2010, I live in the North West Suburbs of London on a smallish Crescent off the A1 and my elderly mother in law and wife were trapped in our house for nearly 3 weeks as the local authority did not have the resources to clear the ice off the road and pavements and grit the road. Can you imagine what sort of service, if any will be available after 2 weeks of poisonous fogs followed by extreme cold, snow and ice? If we were living in Switzerland or Canada where they are used to such inclement weather we might have a chance, however in the UK, forget it. It will (in my humble estimation) be totally down to the individual to organise, heat, light, power, food and maybe gas and water. Lets for the sake of argument consider the worst case scenario, if it doesn't get this bad then we can all breathe a collective sigh of relief.

Day to day living after the Laki Haze

So its freezing cold, you have no mains power or gas or water, could things get any worse, yes, yes they could, as the postman has just struggled up the snowy path to your house and delivered a summons for non payment of council tax. Joking apart, this situation will not be very merry for the elderly, ill, infirm or young children. You should try and keep elderly and Infirm relatives warmly tucked up in bed under 3 or more blankets as elderly people are more susceptible to extreme cold than middle aged or young people. If they insist on coming downstairs you will have to ensure that they are kept warm. Hot drinks would be a good idea, things like "hot Ribena", "lemon tea" (obviously there won't be any fresh milk for some time so start getting a taste for black tea, lemon tea or black coffee).

For an average lunch for four people, you could cook some beans (2 cans) and maybe mix a can of sweet corn in with it. You could open a can of pickled Gherkins to eat with them and a few Carr's water biscuits (crackers) and maybe a couple of apples from the apple barrel, washed down with some cups of piping hot Ribena, life doesn't sound too bad, hot Ribena and not having to go to work for a few months, I would feel positively cheerful on a Monday morning with that thought in mind.

If you have bought your "clockwork" radio you could then listen to the news broadcasts about what is happening around the world and the political repercussions of the volcanic eruption, stock markets would have crashed as businesses failed due to lack of staff coming into the office. I imagine the atmosphere being much like England during the Blitz with a spirit amongst people of helpfulness towards each other and a caring attitude, hopefully neighbours would start caring about the little old lady living on her own next door or the old man with a walking stick who lives at the end of the road. Without this sort of Dunkirk Spirit a lot of the elderly and sick will probably perish in the cold and dark and on their own. Thankfully we British have excellent voluntary services with people like St John's Ambulance Services, The Woman's Institute, The British Legion, Rotary Clubs, Masonic Charities, Special Police Force and many other charitable institutions which I am sure will roll into action to help the old, infirm and young mothers with children who are on their own during this forthcoming crisis.

A few of my readers will remember the winter of 1947 when the UK and parts of Europe were gripped by a terrible freezing cold winter with power cuts, coal shortages, blizzards and frozen streets and pavements. The situation was pretty dire and many small villages and hamlets across the country were cut off for weeks at a time due to drifting snow and ice. We all need to ensure that we have thermal underwear, woollen socks and gloves and other basic clothing maybe some extra blankets and definitely plenty of those campingaz heater gas refills. As mentioned earlier, the main problem will firstly to ensure that supply routes to power stations are kept open and that the key operatives are on standby and the police and army know where they live in order to get them back to run the power stations. Then next in my list of priorities would be to keep the main motorways and trunk roads open so as to enable petrol tankers, food deliveries and emergency vehicles to get from ports and distribution centres to Government supply and distribution areas.

As mentioned earlier I doubt whether local authorities will have the resources or the manpower to actually get teams of gritters and snowploughs out to clean up and open small side roads and crescents (like my street), we would in some ways have to really fend for ourselves for some time.

For those of you who haven't, like I stupidly did, remove my old working fire place and I knew it was a big mistake the moment it happened and I also received a shock when the builder gave me the bill for the removal, I blame the wife for that one, you could (if you had first had the chimney flue cleared) collect fire wood from a park or heath and light a nice old fashioned Victorian blazing fire. If you do not have a clean chimney flue it would be very dangerous to start a fire in a fire place, the flue could be blocked which would mean that the fumes would come back into the room and fill it with dangerous smoke and fumes. If you have stored Campingaz canisters and Campingaz heaters, then you can use them in the room safely although the room should be ventilated.

If the Government has organised heating in the schools and community centres it might be sensible for some of the family (if the roads and pavements are ice free and passable or you have moon boots) to spend some of the day in the warmth of these day centres.

Activity is good for the circulation, tasks such as going out and chopping wood or walking to see a friend would be good for the body (apart from the frail and elderly). Obviously I cannot speculate after the Haze has gone how quickly transport, gas, water and other services are restored, it may be that some companies do restart and shops eventually re open and slowly civilization starts to return. However as mentioned before there will be huge bottlenecks in the transport and supply systems, ships will be waiting to unload, trains and lorries won't be at the right place and maybe the power is still intermittent especially if the winter is long and cold which it definitely will be after a large eruption.

The other main problem will be how you deal with neighbours in your block and in the street, who may smell the cooking or notice that you still look healthy and not emaciated. You have several options as to what you can do about this issue.

Firstly, don't overeat, keep yourselves on fairly strict rations so you do not put any weight on, it will probably be healthier to lose a little weight and the neighbours won't suspect that you have a hidden cache of food.

If you do start putting on weight it will definitely be a sign that you have food hidden away. I am sure that people will have some food anyway and the Government will probably be handing out subsistence supplies of water, bread, some canned foods and freeze dried army style rations, so the odd cooking smell coming from the house shouldn't provoke a major riot. Cooking a few beans and sweet corn in a house or flat with the UPVC or wooden windows shut wouldn't create any real smells or smoke anyway. If you decided to cook some sort of curry, chilli or paella or something with strong cooking smells then you are asking for trouble. People living in converted flats would probably have more of a problem here than people living in a purpose built flat or a house.

Secondly, if you do decide to include the neighbours either side of your house or flat in on the fact that you do have a supply of food and cooking equipment, make sure you have a long chat with them firstly to make sure you can trust them and that they don't in turn start telling their friends and family, as soon you will have many mouths to feed and a dwindling supply of food and drink to go round. Be firm, make it clear to the neighbours that if you help them it is only on the basis that they keep their mouths shut and if you hear that they have been blabbing then you will cut off or restrict their food supply. Unless you trust them 100 % I would advise keeping your provision supply secret. I would advise splitting your food supply as mentioned before, so that if the local neighbours burst in and ransack your cache of food you have a decent backup supply hidden elsewhere in a garage away from your home or in a relative's house etc, don't put all your eggs in one basket. I doubt personally that the situation will be so dire that neighbours come to blows over a can of beans or bottle of water, my thoughts are that the authorities will start distribution fairly quickly of basic foodstuffs and that we British are not going to turn on each other in the midst of all the hardships going on around us, I am sure that people will share out what supplies they have and that we will all be in the same boat as it were and that common sense and decency prevails. We must definitely ensure that we help elderly or housebound or mentally ill people in our own streets and neighbourhood as they simply won't survive without our help.

I am hoping that firstly enough people will buy my book and therefore most people will have some supplies, secondly the Government will have to do what it can to try and issue emergency rations from local distribution points and thirdly eventually, maybe in fits and starts after a couple of months after the fog has dissipated some supermarkets will be attempting to get supplies running back to their stores, the public though when they hear of a delivery will probably descend on the supermarket and clear it out fairly swiftly after any delivery. This will go on as long as people consider that rationing of food and limitation of supplies is continuing.

Just like WWII, I am sure that a black market will quickly spring up whereby shady characters will be hanging around on street corners (after the fog has lifted) who will for a price be able to get you a steak or fresh bread or anything else from caviar to "pate de foie gras" and some "Bolly". It may be that cash is in short supply, the banks will probably be closed, cash machines will not work or may have been looted, it may be a case of barter, you will have to give the wife's diamond engagement ring to a shady character in return for a side of beef, or a side of smoked salmon. Throughout history from ancient times to the fall of Berlin in 1945 there has been black market profiteers, who have made fortunes out of other peoples hardships and misery. This unfortunately is the way of the world. My system of food collecting before an eruption will of course help you to hang onto the family silver as you will not need to deal with these unsavoury characters that is, unless you are desperate for fresh fried eggs, bacon and toast.

I foresee that an extremely cold winter will probably carry off many more people than the fog, there will be many vulnerable elderly, ill and infirm people whose carers will simply not be able to turn up for work or will run away. My advice would be firstly if you have elderly or infirm relatives go and collect them as soon as the news breaks of a major Icelandic or super volcano going off. Alternatively if you cannot look after them try and get them into a private or state run home, the Government will try and prioritise these old age homes for supplies of food and heating oil, it will probably be their only hope of survival if things look bleak. As stated earlier try and collect as much medications for elderly relatives as you can as pharmacies will be prime targets for looters once things start to happen. I would hope that the Government in league with the NHS has some basic contingency plan to try and collect the elderly or infirm and those with mental problems and get them collected by ambulance and put into nursing homes or hospitals whilst the emergency and freezing winter continues.

If anyone falls seriously ill, you will of course probably have to try and get them to hospital yourself, that is why I recommend keeping a few car spares and spare car battery on standby and a full tank of fuel. Again I would imagine that the Government would try and prioritise hospitals for heating, power and food and it would probably be your only chance of saving someone should they fall seriously ill. The Government would probably have the army out patrolling the streets of all town centres, hospitals, stations, ports and airports to try and keep a grip on matters. Troops would probably (if things got really bad) shoot any looters on sight, this sadly would be the only real way in this sort of emergency to keep the country from falling into anarchy. The UK historically is not noted for its rioting subjects, we have had a few, The Miners Strike of 1981, the Poll Tax riots of the early 1980's, going back in time the Gordon Riots of the Eighteenth Century were the only major disturbances in the UK. Whilst of course, on the continent there have been many serious riots over the centuries, most notable being the French Revolution of 1789 (possibly caused by the Laki Eruption of 1783), 1848, the year of revolution across Europe, The Russian Revolution of 1917, revolutions in Berlin in 1919 and many more too numerous to note. The point I am making is that I do not think that the British public will riot as has happened in other countries at other times, there will be a few isolated examples of looting of food shops and supermarkets, however I feel that we will all pull together and the Blitz and Dunkirk spirit will prevail and that neighbours will keep an eye out for each other and people who have food will I am sure share some of it with those of us whom are less fortunate.

If the following summer is like 1784, 1785, 1786 and 1816 years without summers due to major volcanic eruptions, then the failure of crops in the Northern Hemisphere will pose further major problems for governments around the world, if there are no vegetables or wheat crops due to continuing frosts throughout the following summer months after the eruption, then although I have planned for basic survival for up to about 9 months it will be difficult for all of us to keep going into the winter after the eruption without fresh food. I am hoping that countries in Asia, Australia, New Zealand and South America will make up the shortfall in crops such as rice, wheat, barley, oats, fresh fruit, meats, poultry and other foods.

As in Iceland after 1783, although many of the citizens were starving in the interior of the country, they were still exporting dried and salted fish to the USA and Europe at the same time as there was no organised distribution network or relief organisation. The seas will have to be fished around the UK to feed the inhabitants, although I know that currently some of these fishing grounds have been closed due to overfishing by some of the EU fleets we will have no choice if we are going to feed our population apart from fishing intensively to keep everyone going. There could well be military confrontations over fishing rights and food distribution and water, if things get serious, as every country will be trying to feed and water its own population, probably at the expense of its neighbours. Whether the EU and the Euro survive in these conditions is a matter of conjecture, I personally do not see it surviving as populations without enough food or water will pressurise their governments into action to protect those populations.

Perhaps the Government will rethink its plan to scrap the Ark Royal and other ships of the Royal Navy and reduce its budget, we will probably need the help of the Navy, Army and RAF (Royal Air Force) for evacuations, search and rescue missions and possibly patrolling our coastal areas to protect our fishing fleets after a Laki style eruption, we would look very silly if we have a Navy, Army and RAF with no teeth, whilst the rest of Europe is busy plundering the North Sea and Irish Sea with impunity to feed their subjects. I am sure it won't come to blows, however without the real clout of an aircraft carrier and major fleet it will be hard work negotiating from a position of strength when in reality we would have made ourselves much weaker than we need be.

CHAPTER 4

Surviving a Super Volcano

So far we have merely been discussing what would happen if a medium sized volcanic eruption in Iceland occurred, these eruptions are fairly small compared to what a super volcanic eruption could do. The last major super volcano to partly go off was Tambora in Indonesia in 1816, even this eruption although it was a fairly major eruption was only a partial eruption, this caused ash clouds to cover the surface of the Globe and there was no summer for about 2 years. However it has been deduced that this Tambora eruption was not on the same scale as the Tambora eruption of 75,000 years ago that produced molten hot spheroids and increasing the air temperature so that many forests and grasses self ignited burning great swathes of the planet, followed by acidic sulphuric acid rain and probably a nuclear style winter which could have lasted for maybe 50 years or more. This event it has been proposed pushed the human race to the edge of extinction and probably only left about 5,000 humans on the planet, it nearly wiped us out. This was probably the most dangerous moment for the human race since we walked upright about 5 million years ago.

If a "Yellowstone style super volcano" erupted, the explosion will be massive and would blow thousands of tons of ejecta material into the upper stratosphere, the huge magma chamber under the caldera holds about 1,500 to 2,000 cubic kilometres of magma which is eruptible. The ash cloud will cover the globe, anyone within about 350 miles radius of the initial blast will be incinerated by magma bombs, pyroclastic flows and a huge blast wave. Huge amounts of ash will then fall, even as far away as the east and west coasts of the USA, all air travel will stop, many power grids will collapse as ash short circuits power lines and the weight of ash buckles pylons and a nuclear style winter will soon follow once the sun is blocked out by the ash cloud. The next thing that would happen about 3 hours later is that molten hot spheroids will then start to fall to earth from Germany, France, UK, Ireland, USA and Canada and maybe even Mexico. These spheroids will set fire to wooden structures, woods and farmlands and some cities where there is much wood in the construction will go up in flames, maybe some of the wooden medieval cities of France Germany and many of the wooden built houses of the USA will go up in flames, these spheroids act much like "incendiary bombs" did in the last war when German and Allied Bombers rained down these highly combustible bomblets on major cities causing fire storms in cities such as Hamburg, Dresden and to some degree Coventry and Plymouth and the Docklands of London.

You will have to imagine that this scenario of molten hot bomblets were falling over much of the USA, Mexico, Western Europe, Scandinavia and Canada and parts of Eastern Russia, the emergency services would be overstretched to bursting point, there would unfortunately be no one coming in a hurry to put out your house fire, take you to hospital (which might also be on fire) therefore you would have to put out the fires on your own when opportunity arose. I would imagine (although I have not tested this theory yet) that these spheroids would possibly not set fire to houses with slate or tiled roofs or even those roofs which are flat like the roofs of factories or some blocks of flats. They might melt pools of roofing asphalt (which could later be repaired), however I also fear that some bigger chunks of Ejecta would also be flung up into the Stratosphere and if melon or grapefruit sized flaming bombs of magma start landing on your tiled or slate roof, there is a good chance it will smash through the roof and set fire to the combustible wood and stored materials in the attics of houses. Houses and blocks of flats with concrete style roofs would probably fare better. However I doubt in the UK if there would be too many larger style bombs, I think that any spheroids which land in the UK will not be that many as the distance to Yellowstone (or Tambora) is over 4000 miles away.

The atmosphere for the few weeks after and during the eruption will also be temporarily much hotter than normal, it is possible that dry vegetation could self combust due to the potential heat, this will have an adverse affect on the health of the old, infirm and small children and the obese as these members of society are more susceptible to temperature changes than other members of society.

After these two disasters, spheroids and heat will come an acid rain of mild sulphuric acid which could cause skin and face burns to any humans who are exposed to it, many cattle, sheep, pigs and goats and other farm animals and wild animals will be killed by the acid rain, forest fires, spheroids, blast waves, pyroclastic flows, bombs and then after all this the gray skies will get darker, the sun will not shine for about 2 years and a nuclear winter will ensue with temperatures dropping down to − 30 deg C at times. This will not be a pleasant world to live in, places like the Amazon Jungle and the Equatorial African Jungle will start to die, many crops will fail across the world, wheat in the USA will not grow, wheat, barley, oats and fruit will not grow in Europe and Russia and the rice crop may fail in China and S E Asia.

Those who are resourceful, and who have read my book and listened to its lessons will hopefully have a better chance to survive, society which today we take for granted with TV, broadband, electricity, gas, tap water and all the other veneers of civilization would be pushed to the very edge of its ability to survive this catastrophe and rebuild after this disaster. Governments around the world would be reduced to a few hundred men and women cowering in some cold war bunker trying to organise emergency services that for the most part wouldn't actually exist anymore, the army and police (if they survived in strong concrete reinforced barracks or underground bunkers) would have to stop looters ransacking any shops, offices or undamaged buildings. They would try and organise relief centres for those that survived this holocaust at hastily partially repaired with tarpaulins, community centre, schools and maybe some church's, cathedrals and public libraries and similar sorts of buildings where the governments would try and organise feeding centres and pour their resources into keeping these nuclei alive.

Thousands and thousands of humans will die, there will be no chance of saving many people from the severe violent natural disasters described above. I would imagine most people especially those living hand to mouth or in central city areas would be hardest hit, those in rural communities might be able to survive for a while on grain and rice stocks and other silos of food stocks that hadn't been sent for processing yet. People living near the coast might still after a month or so start fishing again and temporarily fish might be the main staple for most people near coasts, however fish feed off smaller fish and small animals called krill and prawn like animals, these feed on phyto planktons, however phyto planktons which are the base of the marine food chain survive due to sunlight, if no sunlight reaches them for 2 years then many types of larger crustaceans and fish and mammals, including dolphins, sea lions and whales will start to die out. I do not think the ash clouds will cover every inch of the globe as winds continue to blow and it may be that remote areas of the oceans or in Antarctica that these planktons do survive and keep some of the food chain going. After all these supervolanos have blown their tops with depressing regularity probably somewhere in the world every few thousand years or so, some like Tambora about 70,000 years ago were mega super volcanoes, whereby, even more ejecta and ash than is normal erupted. However as we are here now and I am writing this book that is proof that mother nature and us humans do seem to be able to regroup and bounce back after suffering terrible times.

I would state that the unfortunate element of modern day lifestyles is that everyone has become a specialist to such a degree that for example whilst a radiographer might do a fantastic job of producing x rays of a fracture of the tibia and be able to work the complex controls of a multi million pound cat scan machine, which looks like it's come right out of a science fiction movie, that same person, if you asked them to change your flat tyre at midnight on the Hendon Way in North West London wouldn't have clue and would have to ring the AA or RAC to change the tyre. How will our x ray radiographer cope, if after running out of food in the city, he or she manages to get to a farm near St Albans in Hertfordshire and then has to help farm using medieval implements, hoes, rakes, shovels and hand operated machinery to plant sow, reap, glean any vegetables, root crops or other crops that this commune decide to try and grow. This x ray specialist will have to very quickly learn a complete new set of skills to survive in this "Brave New World" where completely different sets of skills and living styles might have to be learned.

Sitting in Starbucks writing Harry Potter Novels will, for a few years at least, be a thing of the past as will sitting on humid, stuffy underground trains stuck in tunnels, so things won't all be bad. People will have to get back to basics and learn to live in sync with the new reality that a harsher less forgiving nature has taken hold of the planet at least for a few years.

Where does all this leave our house dweller that has taken all my precautions and has bottled water, cans of food, packets of dried food, campingaz canisters (kept at a cool temperature to keep them safe). Life will initially be a lot harsher than if a major Icelandic volcano blows its top, especially with regards to the spheroids rattling off your tiled or slate roof and possibly setting garden hedges and bushes and garden sheds on fire, then there will be the acid rains and heat wave that will have to be survived, however if careful precautions are taken as described in earlier chapters then I see no reason why life wouldn't eventually go on as normal. After about 3 months if the Government can organise some resources, I am hoping they will prioritise water and also sewage services. The reservoirs should be able to cope with the acid rain as this will dilute the acid and the water companies can then put in alkaline additives to neutralise the ph factor of the water which should then be potable (drinkable).

However the infrastructure that we are used to from shopping centres, fire stations, libraries, town halls to bus stations and post boxes will all have changed. It may be that the Government tries to get all survivors into community centres and schools, similar to the Hurricane Katrina Centres at the Rosebowl. This will help the authorities to keep tabs on everyone and help with the distribution of state aid, food, medical help, water distribution and all the other matters that the state will try and provide, however there will be inherent risks in living in this kind of loose "camp" such as diseases, typhoid, cholera, malaria or any other diseases which vermin or people living in close proximity can carry, and if the sewers break down and water supplies get contaminated further health and sanitation problems will occur. Heating and lighting might also be intermittent depending on how many supplies the Government manage to get through to each centre and if any foreign aid arrives.

As in New Orleans things became violent in the Rosebowl and there were fights amongst the public and no proper sanitation or co ordination and this is when the rest of the USA was ok, we are talking of a situation where all semblance of normality across the country has collapsed and every region or area will have to fend for itself apart from sporadic help from central government. There will probably be emergency broadcasts of where to go for relief supplies, food, water and medical help on TV and radio and from loudspeaker cars and vans as if there is no power how will you know where the distribution centres are or when a delivery of fresh food has arrived in your locality.

Personally I would rather sit it out in my own home stocked with most of life's essentials, using my wind up clockwork radio to listen to bulletins or occasionally connecting one of my car batteries to the radio or TV (through a dc converter of course) to listen out for news. Hopefully many of your neighbours along the street or in your block will also have been reading these pages and been saving food and supplies for a rainy day and when that day came they will not have to trudge off to some communal farm in St Albans or go to the local school to be bullied or harassed by gangs of thugs or ruffians. You will be able to control your own destiny and hopefully after one to two years things will start to get back to normal.

Once the skies clear and proper crops can be planted, bread production will take off again, tankers will be able to dock at Milford Haven and Dartford to unload petrol and diesel again and "Britain's Got Talent" and re runs of "I'm a celebrity get me out of here" will start to be shown on TV again. Council and Central Government employees will start returning to work, rebuilding of damaged public and private sector buildings will slowly get underway, it will be much like London after WWII when although much of the capital was still blitzed and bomb sites were a regular feature of life in the UK, people did start to return to work and life once more started to return to peacetime situations. Train services and the Underground will start up again and hopefully life will start returning to normal once more, but will it?

CHAPTER 5

Surviving the Aftermath
of a Volcanic Eruption

So, weeks have gone by since the Icelandic Laki eruption, you have survived the spheroids, the poisonous gases, the power cuts, fuel cuts, the empty shops and supermarkets. You and your neighbours have become a tight knit bunch of hardy survivors, you have taken it in turns to queue at the local distribution centres for the meagre Government handouts of bread, water, canned foods, butter and oils. However after a month or two things start to look more normal what will everyone be doing next, how will the wheels of industry start to get turning again?

I would imagine in the immediate aftermath of the eruption (Icelandic, not super volcanic for this exercise) the stock markets would probably have gone into freefall as everyone tried to cash in on equities to become liquid whilst instability grips the markets, shares would have plummeted in most categories, probably apart from food stuffs, commodities and oil and power which in the short term would probably have risen, however once the full scale of the damage was known they too would probably drop like a stone, because although oil would be in demand there would be no way to get it from the Middle East to the end users in the UK, France and Germany, how many lorry drivers would want to drive a huge petrol tanker through a poisonous yellow fog where they couldn't see where they were going, this goes for foodstuffs and other commodities, the demand will be there but the ability to get these supplies to where they will be needed will cause an insurmountable problem for these producers. However gold and precious metals would have gone through the roof probably hitting £ 1,800 per ounce easily (for gold). Everyone would be concentrating on looking after their families and property so it is doubtful whether anyone would struggle through poisonous fogs and falling ash and acid rains to get to their city or West End jobs, transport would probably grind to a halt pretty quickly too, buses, trains and taxis would not be running. Most people would either be at home, hoping to sit out the eruption or foraging for scarce food, water and petrol supplies, which would run out within days if not hours.

What good will money be or credit cards if there are no goods and services to buy. If you go down to Tesco to buy some potatoes and all the shelves are bare, with no re supply in sight, what are you going to do with your money, there will be no one who wants money at this point, it will be a useless commodity, therefore for a while society will be self levelling, the greatest Russian Oligarch will be queuing for his loaf of bread with the East End market trader, in fact the market trader will probably have a big advantage over the oligarch, as he will know how to duck and dive and will quickly understand that bartering goods for services or other goods will be the only real commodity of value. If you have a Cartier gold watch studded with diamonds worth £ 50,000 but you and your family are starving hungry then today's value of a £ 50,000 Cartier Watch is three carrots and a can of Heinz baked beans on the black market. People will have to start bartering for each other for goods and services, a tailor will sew the hole shut in your trousers for 2 cans of sweet corn, and the cobbler will repair your shoes with holes in them for 3 apples and 2 bottles of mineral water.

I am not saying that this situation will last forever, slowly as society gets rolling again and transport picks up and people can get back to their jobs and most importantly that shops, supermarkets and suppliers start getting regular supplies the old system will start creaking back into life. However for maybe a few months those who listened to what I am saying in this book will have a very valuable hoard of valuable assets, I am not saying to go down to Bishops Avenue and start selling Russian Oligarchs cans of beans for diamond tiaras, however the balance of power will have changed, whomever has any supplies will be king for a period of time.

There will be considerable problems for central banks and high street banks and building societies as no one will have been able or possibly had the money to pay their mortgages, household bills and credit card bills. Who knows how central government will be able to kick start the economy and at what parity the pound will start up again against the euro and the dollar and the yen. Initially as the value of goods and services is restricted hyper inflation could set in, that means that 5 lbs of potatoes at Tesco which is today £ 2.50 (I actually bought two huge sacks of potatoes for £ 2 last week, some prices are ridiculously too low currently) would after the eruption (if there were any left in the supermarket) would suddenly go up to £ 40.00, then when no more food was available in the stores, the black market price of the potatoes could be £ 200.00, it would be up to the seller of the potatoes to set whatever price he or she thought they would be able to get for the products, then the black marketer would have to go out of London to the farmer and pay him probably £ 100.00 for the next bag of potatoes and so it goes on, a week later those same potatoes would be worth £ 1,000 and at that point money becomes worthless as most people after the first few days of trying to buy food on the black market would have simply run out of money, so temporarily the oligarch is in the lead, however when he has to pay £ 1,000,000 per bag of potatoes then even the oligarch will have trouble and he will sink down to join the rest of us.

Therefore for those of you who listened to my advice, you would have a house full of commodities that everyone will be clamouring for and for a short period of a few months you could make enough money to retire on, however do not sell too much of your stockpile as then you won't be able to survive the lean months before agriculture, industry and shipping supplies starts to re emerge. As previously discussed, and as can be seen above you will have to be very circumspect regarding whom you tell that you have some hidden supplies, burglars, muggers and all sorts of desperate people will be hanging around, begging, stealing and worse to keep themselves fed. The Government will have troops patrolling streets and curfews in place to stop looting however cities are big places and if the petrol runs out, which it will fairly soon, the authorities will have a hard time patrolling and controlling large areas of inner city areas. With regards to country areas I think that there will be even less Government control as there will be no mobile or land line phone network properly operating and I fear that small rural communities will have to take the law into their own hands for a while and organise vigilante patrols to ensure the safety of villagers and towns folk from marauding gangs from the inner cities who have come out into the country side to try and get some food, from farms and villages. This is why I am encouraging streets to form some sort of neighbourhood committees to communally distribute food on a street by street level or block by block or estate by estate, and this goes for security, patrolling your small patch of land will be very important to protect your neighbours from outside attack and pilfering.

The Government should really now start issuing guidelines on how to organise such committees and communes for small communities and streets around the country. In Switzerland as I mentioned earlier, since 1962 all properties built not only have to have a safe nuclear proof room, also there must be a store of food, water, air and water filters and other standard survival equipment by law.

I was for a while in the 1970's in the Combined Cadet Force whilst at a public school in Surrey and whilst in the army cadets we carried out weekend camp and manoeuvres at Aldershot, when it came to the evening meal we were given cans of sausages and corned beef, I looked at the date stamp on the bottom of the can and it stated 1941, I would have to say that as we were hungry we cooked those sausages and corned beef and it tasted fine and there were no after effects. I have been informed that the Government might have sausage and corned beef mountains hidden in secret bunkers around the country, if so I am sure these will be distributed in time of emergency to a grateful if hungry nation.(let's hope the sewage works don't pack up too soon afterwards).

Returning for a while to the situation after a few months down the line, I feel that many small firms and businesses will have a very difficult time picking up the pieces after this catastrophe as there will be various old bills to pay, even if some government taxes and vat is wiped away, the whole cycle of commerce will have been disrupted and many firms will shed staff so there could be several million more unemployed people in the UK who will also need helping with food, housing benefits, council tax relief and other benefits at a time when the Governments finances will be in a terrible mess. The only consolation will be that most other developed nations in the Northern Hemisphere will be in the same boat.

If we take Germany after its defeat after the first world war, when hyper inflation set in and people were taking wheelbarrows of reichmarks to the bakers to purchase a loaf of bread, like Zimbabwe who printed a 500,000,000,000 note which was worth about $ 100.00 the Germans actually invented a new figure called a milliard, which from memory was worth 1,000,000,000 or thereabouts. The problem with hyper inflation and massive unemployment and poverty is that it is a breeding ground for nationalism and the spread of fascism, parties like the BNP will be shouting England for the English and we cannot feed all these foreigners, this is why everyone must pull together much like the blitz war years and keep far right and far left extremists out of politics and keep the mainstream parties in power. If one forgets history one is doomed to repeat history. All countries around the globe will have to help each other whether it's through the United Nations or the IMF or some other new organisation, is not important, what is important is that we all realise that disasters such as these Icelandic and super volcanoes are a regular feature of our planet and we will all have to stand together to try and rebuild our society whether it be financially or physically, by all pulling together.

Eventually life will come back to normal, firms will start to re hire staff, shoppers will wander down their local high street in search of that elusive bargain. Trains, buses and plane's will all start to go back to their normal schedule's although the collateral damage to people will have been terrible, some will not have survived, some will be very ill from the gases and others from inhaling ash. Governments will be in a terrible financial mess as they will have to fund massive repair bills for all the damage done to houses, schools, government buildings and trying to help uninsured householders and blocks of flats and council blocks all damaged by the eruption and its aftermath. The Government will have very little revenue coming in as many firms will be closed or on short time working. The Government will have great difficulty in raising money on the international markets as all countries and investors will be in the same boat and the IMF will be overstretched to the point where it will not have any funds to give to anyone, all that central banks could do in this situation is to print more money and then hyper inflation will set in again and everyone's savings and investments will be worth very little. This will be a time of wild financial fluctuations with stock markets rising and falling in wild gyrations as rumour and gossip drives the world economy. It may be that the dollar or the Chinese Yuan is used as a currency of last resort where other currencies are pegged to it.

This period of perhaps several years will also be fraught with difficulties, as although the central government will probably be able to organise basic distribution of staple foods to the public like, bread, potatoes, tea, sugar butter, some meat and chicken, the currency, i.e. the pound may have collapsed again and would probably be hundreds of pounds to a dollar or Chinese Yuan. The Government might bring back rationing as during WWII, when everyone was given a ration card and you could only buy so much meat, bread, milk etc per week. The Government will issue some sort of stamps to stick in the ration cards, that will authorise the holder to a certain amount of food per week, depending on the size of the family. The bigger problem probably won't be food production and distribution which will probably be the Governments top priority after the eruption, the main problem will be one of kick starting the world economy back onto a peacetime footing and working out exchange rates between countries and possibly barter placed deals might happen in the short term, whereby Russia turns the gas back on for 2,000 sheep per month from the UK, or some such deal.

The UK is still sitting on a huge amount of coal hidden deep underground and although many of the mines have been flooded or allowed to go to a poor state of repair, with some hard work, powerful pumps can pump the mines dry, miners can then go down and shore up the mines and soon coal will be produced again and we won't be dependent on Russian gas and Middle Eastern Oil. The earth's atmosphere will already be full of gases and ash from the eruption which may take several years to disperse and although I am not advocating returning to a fossil fuel economy we may be able to burn coal cleanly by using electrical particle catchers placed on the power station chimneys. For a few years we might only have coal to rely on and whatever food and crops we can grow ourselves and the generosity of the EU and the USA, our allies. I know that I keep harping back to how we all survived the war, however that is probably the closest similarity that I can think of, it was a time when a human enemy was bombing our towns, cities, train stations, dockyards and factories on a regular basis, as well as regular attacks on civilian as well as military and industrial targets. We all suffered together, huddled in underground stations (which actually could be used again this time to protect and feed civilians, they would be a good place to store food and water by the Government) and in our nissan shelters with our buckets of water and stirrup pumps to put out any German incendiary bombs that might fall on the house.

The Government then had slogans like "Dig for Victory" which was a very good idea, most public parks including Hyde Park and Regents Park were turned into allotments for growing peas, carrots, potatoes, courgettes, cauliflowers, beetroots and all other manner of fruits and vegetables. Although this campaign was partly done because the German U Boats were sinking most of the food supplies that were coming across the Atlantic Ocean, However if the UK has no cash to buy food stuffs and other materials like concrete and steel, then all we can do is barter other goods for these goods or grow our own foods and dig our own steel ore and coal out of the mines. We will also have to start manufacturing again, producing items that other people in other countries want to buy so we could trade, barter or sell our goods to other countries to try and start to rebuild our balance of payments. During the last war there were very efficient ministries organising the nationalisation (during the war) of farms, mines, trains, shipping, air freight and all other aspects of productivity so that efficiency rose and the UK produced much more during this time per capita than before or after the war. True there were also strikes and workers unrest, however on the whole everyone worked for the common good, apart from the black market racketeers and the seedy underworld of prostitution, drugs and other nefarious wheeler dealings that is part of most societies.

As previously discussed, much depends on the severity of the eruption and the proximity to the UK, although the Icelandic Volcanoes are smaller than the Yellowstone Super Volcanoes, they are much closer, so strangely the UK, if one of the big Icelandic Volcanoes blows, is probably going to suffer nearly as much as if a super volcano blows 4,500 miles away. Another interesting feature of the UK's proximity to Iceland and its nearness to the USA is that with the Icelandic eruptions we might (if the prevailing winds are South Westerly's) get inundated with ash and poisonous gases within a couple of days of the eruption, and although the super volcano is a much bigger and many times more dangerous than the Icelandic Volcanoes, we would also receive a much longer warning period, with regards to the super volcanoes as, if Yellowstone goes off on a Monday even with prevailing Westerly's (winds blowing from the West as they normally do) we would not see much evidence of ash clouds blotting out the sun until Friday or Saturday, giving the population at least some time to make arrangements, travel, try and get some food, water and clothing before the full force of the eruption hits our shores. Of course there will potentially be showers of spheroids and other heat related phenomenon which would hit the UK and Europe within 72 hours from eruption.

Chapter 6

What to do if you are abroad when the eruption occurs

Recently when the Icelandic Volcano EJ blew its top, hundreds of thousands of people were trapped all over the world being forced (in some cases) to actually spend a further five weeks in the Bahamas, (all expenses paid) before being able to fly back to Blighty, what a terrible world we live in being forced to lie on the beach drinking endless vodka martinis and reading all the latest Clive Cusller and Jeffrey Archer books and eating fantastic six course meals before retiring to your beautifully appointed "presidential suite" for a peaceful night's sleep, dozing lightly then listening to the sound of the rolling waves on the beach crashing onto the shore falling into a deep untroubled sleep. Obviously not everyone was trapped in idyllic locations, some had to put up with five weeks at The Waldorf Astoria in Manhattan, whilst some were in Dubai, Thailand, Sidney Australia or merely in Calais for a weekends booze cruising (they, though probably did get home as the ferries continued to operate).

I believe this eruption of Eyjafjallajokull was a warning shot across our bows, all air traffic across the Atlantic ceased fairly quickly as well as all flights across most of Europe from Spain to Norway and from parts of Russia to Ireland, what happened to those travellers who were stuck away from home and whose families were desperate for their partners, wives, husbands, children, relatives, to come home, especially important if the bread winner is stuck abroad for some time. There were heroic stories of people paying Italian and Spanish minicabs hundreds if not thousands of pounds to get them to Calais. Others told wondrous tales of catching buses, trains, hitching lifts and all other manner of escapes to get back to Calais. What happened at Calais, everything backed up as everyone, especially those that had been on business abroad paid top dollar to get on Eurostar to get back to London, or elsewhere in the UK. However, for those trapped at Calais trying to get back, it sometimes took several days to get passage on one of the ferries back to the UK.

What about people stranded further afield in Dubai for instance, there is no way on earth that you would want to get into a taxi or a bus with a family of four and luggage and just hope for the best. These people literally had to pray for a miracle, i.e. that the eruption ceases and airspace opening up again and eventually after much phoning around and badgering of travel agents and tour reps most families returned to the UK to tell their story in the News of The World. However let's take a scenario where the volcanic eruption from a Laki sized volcano goes on for weeks and months and belches out thousands of tons of ash cloud that no plane would dare to fly in, covering most of the Northern Hemisphere. There you are, enjoying your well earned holiday with your wife and kids in St Lucia, at the Sandals Regency La Toc (Ex Cunard La Toc Hotel, a very smart 4 star hotel on this tropical island) you turn on CNN and you see the headlines, "huge ash cloud covers the Atlantic and Europe, no flights at all, everything cancelled for the foreseeable future". This is, for arguments sake on a Friday and you are meant to be boarding your plane at St Lucia's airport at Castries, the Capital of St Lucia on the Monday morning at 7am. So you quickly go and talk to the tour rep, he (or she) will quickly get many gray hairs as he tries to calm down hundreds of nervous, worried and increasingly angry customers, and all he / she can say is, "I've spoken to head office, there is nothing anyone can do about things currently so we are arranging for you all to stay in the hotel at no extra charge to you until further notice." As a point of fact in some smaller hotels and independent hotels, the proprietors demanded that all guests pay them cash or credit card for the extra accommodation, although I understand that the airlines then re imbursed most of the monies back to the guests at a later stage. So what do you do, no planes, probably only some small fishing boats on the island and some small private single or twin engine aircraft at the small Island Airport?

Firstly using your laptop, BlackBerry or mobile, you call the office and try and re schedule some of the critical big meetings, sales pitches, important client meetings that were due to take place the next week, you find out that other key members of staff are also trapped abroad and the MD is fuming and not to be disturbed if you want to keep your job. You can plead all you want that it's not your fault and it's an act of god, what can you do about it. Suddenly your carefree sunny holiday that you have worked all year for is turning into a nightmare, the happy Hawaiian shirted guests are all huddling together trying to think of ways to get back to the UK or USA wherever they might live. Wives start to get worried about elderly relatives and friends family back home as pictures of a yellowy fog rolling around the streets of London fill the TV screens. The kids though, think they are in seventh heaven and are having a great time surfing, playing in the waves and generally making their parents more frazzled as they try and plan their escape back home.

The next morning, (Saturday) everyone besieges the hapless rep again, the guy is now probably on more Prozac than is healthy for an adult male in their early 40's. What can this rep possibly do, no flights are available apart from St Lucia maybe to Martinique and other local Island hop flights and then onto Venezuela or Honduras, or even Mexico City, let's call our hero Mike and his wife Laura and the two kids Harry and Lucy. Mike is in a fairly high powered job with a solidly performing insurance company based in Central London, however he has many bills to pay, big mortgage on the Georgian house in Islington and many other liabilities and HP and debts to cover on a monthly basis. The rep, George is doing his best to help, however the airlines liability is to get the client home from their holiday destination back to the UK and if the airline or tour operator tells the client to stay in the hotel (at the tour operator or airline's expense) then if the customer then gets off the Island to another destination (not home to the UK) then the carrier or tour operator is not then responsible for any further costs at that point. However this is still a grey area and I am sure that these issues will be rattling around the UK and EU courts for many years to come. Much of these regulations can be traced back to the Warsaw Convention on Travel from

1929 and as much as Ryanair's Michael O'Leary might like to hope that there was no 1929 Warsaw Convention to protect air travellers, it is still on the statute books across the world.

Many of the guests get together to discuss the meagre options placed their way, the tour rep has managed to arrange (for those who will accept it) an onward flight from St Lucia to Martinique's large International Airport and from there onwards to Mexico City, however George then states that, that is as far as he can work out the return to the UK and that clients will then have to make their own way from Mexico City on to the UK under their own steam and he vaguely hints that the tour operator might pick up the tab for the rest of the journey home to Blighty after some research into what happened, however this is only a vague promise.

Mike speaks to his office on the Monday morning and is told that Barry (the MD) is blowing a fuse about a major meeting coming up in a week's time with a huge client and vaguely threatens that Mike had better be there. Mike is also told that half the staff couldn't get in to work due to some foul smelling fogs that had descended on London and that there was precious little food left in the shops to eat and most of the petrol stations had run out of petrol. The Government were asking people to stay in their homes until the poisonous fog had cleared and stating that only those citizens with urgent business or involved in food or petrol distribution should be out on the streets, and only then with some sort of gas mask or breathing apparatus, things were looking bad and the staff were envious of Mike in his island paradise.

Laura was worried about her father in law, Harold, who lives in a large house in Chalfont St Giles in Buckinghamshire who was a widower and relied on a Filipino Carer to cook his meals and help him around the house. Although she had spoken to him on Friday she couldn't get through on the Monday following, some land lines were showing no dial tones and some of the mobile networks were starting to drop calls and showing unobtainable as the systems were under huge loadings and key members of the IT and technical staff needed to keep the systems going, had failed to turn up for work due to the poison fogs and lack of petrol.

Mike went up to the reception area of the hotel to find George, as usual now, besieged by irate and distraught holiday makers all desperately trying to organise flights or boats off the Island and a route back to the UK or for the Europeans and Americans an alternative route home. "Ok guys, take it easy, I have had some input from head office," "said George" "and although we can't get any of you a direct flight home to the UK from St Lucia and we have no ships or cruise liners that we have any arrangements with, we can offer a way off the Island. However head office wants me to re iterate to you all that our responsibility is to get you all home to the UK or Europe or The States and until we fly you there, we will carry on paying this hotel for you to stay here in the interim. If we can fly some of you to another area in Central America or South America, once you land there you will be on your own and we will not be responsible, immediately for your onward passage to your final destination at this point." George pointed out.

Henry a red cheeked large man from Sussex in his late 50's then said, "what the hell are we, with wives, children and luggage going to do then, if you drop us off somewhere in Central America how are we going to get back home to England, you have a responsibility to us, I will get my lawyers onto this immediately."

"Calm down," said George, "we are not saying that we won't reimburse you and your family later when you get back to England, we cannot from this end organise onward travel and recompense from every resort and hotel in the Caribbean and the States. We will make the arrangements to get those who want to go to the mainland and from there you will have to make your own arrangement to get back to the UK and then send us the travel and accommodation receipts that you picked up on your way home and we will get these settled as quickly as possible, that's what head office are saying, in principle."

"Now you listen to me," said Reggie, a 40 year old senior management consultant from London, "your company has a direct responsibility to get us all back home to the UK or France, Germany wherever we boarded our flight from, you can't just fob us all off with some old tale about getting ourselves home and claiming later. How are we going to get ourselves home, what direction do we have to take to get home? What has the British Government done about getting us home? Why can't you give us some money up front to get us all home with?" said Reggie.

George was getting this treatment from all sides and although Mike was in the same predicament as all the other fellow holiday makers, he still felt a twinge of guilt as poor old George got it in the neck time after time from this crowd of unhappy holiday makers.

"I think we should take the offer of a partial trip back home from Trans Global Holidays and thank George for his hard work and attempts to get us all off the Island, we are all in this together and these sorts of discussions are probably going on all over the world, I for one will accept George's offer, when can my family and myself fly out to Mexico City George?" Mike stated.

George looked visibly relieved that one of his angry clients was going to accept his offer, "I think that Mike and his family's acceptance of Trans Global's very fair offer to get everything moving and the rest of the compensation paid when we all know the final cost of getting everyone home is eminently sensible, in half an hour I will be at my desk sorting out the travel arrangements, anyone else who wants to take advantage of our offer please come and see me then," said George, less harassed than earlier.

George then walked off quickly to avoid further arguments. Reggie piped up "Well suppose we follow George's advice and we accept Trans Global's offer of a flight to Mexico City, what happens then, we will have to fund our own hotel bills and onward flights or travel to where? All Transatlantic flights have been cancelled, there are no ships that we can cadge a lift off, we will have to fly Westwards towards Japan, China or Indonesia at our own expense and have to fend for ourselves, it's not a wonderful thought and then we will still be thousands of miles from home?"

Then Henry said, "Look, it's a dangerous world out there and I feel that we should all try and stick together for as long as possible, we are all trying to get home to the UK, I don't want to be in some flea bitten part of the Orient or Middle East on my own with wife and kids in tow, let's all agree to stick together and try and travel home together, there will be strength in numbers and we can all try and look out for each other." The others mumbled their grudging agreement with Henry and Mike's idea as it seemed to be the only way off their tropical prison, which was how most of them were now feeling about their continued stay on the island.

Mike then said, "Sounds like a good idea to me, everyone will be having to make these sorts of decisions together and it will need all the effort and resources that we all have between us to get home in one piece, let's all go and see George in half an hour and let's get rolling."

So this small band of English and European travellers, about 10 families in all, agree that Trans Global can fly them all with wives, children and luggage to Mexico City via Martinique.

So, on the following day this weary band of travellers with all their luggage are being ferried to St Lucia's small airport and flown in two groups on the small twin engine planes to Martinique, a much larger French island which is actually a French Department. After a gruelling 7 hour wait for a connecting Air France flight to Mexico City which arrived on Tuesday 11th October at 7.50am, they all managed to get rooms in an airport hotel very near the main airport in Mexico City.

They all trooped off to their rooms for some well earned rest and agreed to meet again in the hotel lobby at 3pm, to discuss what to do next.

"I'm shattered," said Mike, as he collapsed onto his bed, "let's get some sleep darling, I'm sure we can get us and the kid's home, we just have to persevere and not take any bull from these tour reps and airline people, night, night."

The phone alarm woke them up at 2.30pm, Mike said "Laura, you and the kids have a bit more rest, I'm meeting Henry, Reggie and Vincent and the others in reception in half an hour to discuss our next move," "OK honey, let me know as soon as you agree anything," said Laura, who was barely awake.

Mike switched on the TV softly and kept the volume low, he tuned into CNN and saw more pictures of the yellow fog covering most of Europe and the UK. "The eruption of the Laki Volcano in Iceland has not ceased, it even appears to be growing," said the news anchorman. "This sulphurous fog is extremely toxic and all across Europe people have been again told to stay at home putting wet towels across door bottoms to stop any of the fog seeping into houses or other buildings. Scientists maintain that this fog contains sulphur dioxide and other chemicals which could attack any fleshy internal tissues of humans and death could occur fairly quickly if you are caught outside in this fog with no face mask." The anchor continued, "In the UK as across the rest of Europe large numbers of cattle, pigs, sheep and chickens and ducks have died due to inhalation of this fog, also many homeless people, gypsies, campers and others caught outside have either died or been taken to hospitals where they are being treated for respiratory poisoning, some people are in a critical condition, no numbers are available currently."

"All travel within the UK and EU has ground to a halt, no aircraft, trains, ships, cars are moving, unless emergency services or with breathing apparatus." The killer Laki fog has spread from Iceland on prevailing S E winds across Scandinavia, UK, Ireland, through France, Germany, Holland, Belgium, Spain, Italy, Eastern Europe as far east, as towns only 40 miles from Moscow and as far south as Istanbul. It is feared that many thousands of elderly or infirm, or people with breathing difficulties may have succumbed to the killer fog so far. Volcanologists have confirmed that Laki is erupting through a covering glacier which is causing a poisonous mix of chemicals that are producing this toxic fog over Europe, they say that this eruption could go on for weeks, if not months, many lung and eye problems are also being treated in hospitals in Iceland which has been virtually cut off from Europe by this catastrophe, however aid and supplies are getting through from Canada and the USA as the prevailing winds currently are blowing the fog and ash cloud on a S E direction over Europe and Scandinavia."

He continued, "food and drinking water are starting to get in short supply across Europe as supermarkets have been cleaned out by worried shoppers and many shops have been looted and are now empty. The UK Government has been meeting in an underground bunker near London constructed during the Cold War to

deal with a Nuclear or Biological attack and they are said
to be working on plans to start food, blankets and water
distribution as soon as it is safe to come out of the
poisonous fog bound areas. We are issuing bulletins such
as this one every half an hour currently." The anchor
continued, "many thousands, maybe hundreds of
thousands of tourists, businessmen and family members
visiting other family members are trapped all over the
world and are trying to get back home to their loved
ones, all governments have instructed their embassy and
consulate staff to assist fellow countrymen where
possible, consulates are handing out food and water and
financial assistance and hotel accommodation where
possible to stranded travellers in a bid to help people
keep together and are trying to sort out alternative
routes back to their countries of origin. The other main
issue for Governments around the world is that if all
these returning travellers do get to Southern Spain or
Turkey, if they actually get into the poison fog areas,
they will also be at high risk from respiratory diseases,
so the UK and French and German Governments are
negotiating with the North African Countries such as
Morocco, Tunisia, Egypt, Israel, Lebanon and Turkey to
set up temporary refugee camps, so that the returning
Europeans can wait in the relative safety of the refugee
camps until the killer smog's disperse, however scientists
cannot guess currently when that might be as the
volcano is continuing to erupt."

"In Iceland itself matters are dire, much of the livestock
has now died due to the poisonous fogs and noxious
gases being emitted by the erupting fissure volcano Laki,
we are also hearing unconfirmed reports that the

poisonous gas cloud for some days has enveloped Reykjavik the capital city of Iceland and we understand that many people died of asphyxiation and respiratory related illnesses, the Government there has declared a state of emergency and all travel apart from the emergency services has been forbidden until the eruption ceases. The Icelandic army is patrolling all towns and handing out bottled water and food, the troops have to wear bio hazard suits to enable them to keep patrolling the streets of towns and villages in Iceland. Other countries have promised humanitarian aid including the USA and UK, however with all flights over the Atlantic and Eastern Canada and Europe as far as Moscow cancelled these supplies are having to be brought in by ship which could take at least a week to reach Iceland at a minimum. It is even reported that many container ships and oil tankers are refusing to enter the fog bound sea approaches to Europe, the UK and Icelandic waters for fear of the poison fog, many ships are anchored using their sea anchors off the Bay of Biscay and around Gibraltar." Mike turned the TV off and took the lift down to the hotel lobby, there he saw Henry, Reggie, Vincent and the other dads, sitting on some sofas around a coffee table discussing matters.

Reggie was saying, "this situation is hopeless, I have already rung five travel agents and they can't get any flight going through to Europe, all I can get, in about a week's time is a flight to Baltimore or Pittsburgh, no further North or East than that. If I want to fly east I would have to wait about two weeks to get a flight to Dakar in West Africa or Cape Town with no connections going anywhere else, this is all looking pretty grim, did you other guys see the CNN reports, even if we made it back to Europe how are we going to get through Europe and then onto the UK, I haven't got a BioHaz whatsit suit or goggles?"

Henry then chimed in, "I have also got the front desk ringing around town and I didn't get much better, some flights to San Diego, L.A or San Francisco, there have also been problems with my VISA card, I tried to order a take away meal and my card was declined, when I got through to VISA emergency phone number which I think rang Mumbai, thank god for Indian call centres, I take it all back, what I used to say about Brit companies going abroad, at least they are still open, they told me that as no one was able to get through to the main data processing centre in the UK to keep the main frames going, part of the system had shut down, therefore they told me that many UK citizens even though they had good credit on their cards were showing declined, as no one was available to re set systems and monitor main frame computers currently, so even if we get a flight, unless someone has thousands of pounds or dollars how are we going to pay for the flights, what a nightmare."

Mike then said, "I was watching the news and things are getting pretty bad back home, we all have elderly family there, so for me, it's very important to get back home." Mike then recapped the news programme to the others who had not listened to the programme and they all agreed that they would stick together and try and work a route out together. Mike then said, "why don't we try and get some help from the British Embassy here in Mexico City, on the news it stated that consulates around the world have been asked to help their countrymen in distress, I think we qualify for that description, why don't we ask the doorman to flag down a taxi and lets go and visit the consulate here, in the centre of town." They all agreed to Mike's suggestion and ten minutes later, after informing their wives where they were going, they all piled into two taxis and were taken through the busy petrol fume filled streets of Mexico City to the British Embassy there, which was situated in a grand old Edwardian building in the smarter part of the City.

Justin Urquhart Smythe was stressed out, he had spent the last three days trying to handle hundreds of stranded Brits who had inundated his embassy for financial help, travel arrangements, medical help, missing members of families, business deals going sour and a thousand other complicated issues that he and his tired and worn out staff were having to deal with. To make matters worse, communications between Mexico City and London were starting to fail. It appeared that as no one was able to get to work in the UK due to the poison fog that BT were now, not able to offer any land line connections, the mobile phone companies also had no staff in their head offices so some networks were no longer operating, or they were operating sporadically. There were still some protected internet and secure communications links still operating with London although he could tell by the replies that only very junior staff with no conception of the situation here on the ground were at their desks and they could offer him very little help or relief from all his troubles. Although he did have some money at the embassy, funding all these weary and disheartened travellers was becoming more and more difficult. He saw Martin Symmonds on the first floor landing, "How are you and all the chaps holding out down there on the ground floor with all those lost souls Martin?"

"Well, if we could get a little sleep now and again it wouldn't be so bad, everyone is pretty exhausted, we have been working sixteen hour days and what with the traffic and general chaos here in Mexico City its affecting morale, I have just put a coach load of elderly Saga holidaymakers in El Imperial Hotel downtown, its hard going though many of them are on medication and we are having a big problem trying to get the translations of the medications in Spanish and the correct dosages sorted out, how are you doing in communications?"

"Its getting tougher and tougher to reach anyone who can make any decisions back home, we are losing most lines, only a couple of the secure lines seem to be holding out," "Ok Justin, I'm going back down now to start work on the next lot of waifs and strays, we have so many downstairs that we have had to sit some outside in the gardens and offer everyone a numbering system to get some order down there."

Martin went back downstairs and returned to his desk in the large rear office of the embassy, large old fashioned ceiling fans were stirring up the still autumn air and the flecks of dust shimmered in the midday heat, which actually to the locals, it being their winter wasn't that hot at 75 deg F. However to Mike, Reggie, Henry, Vincent and the rest who had waited three hours to be seen, they were overheated, weary and tired.

"Good afternoon gentlemen, how can we help you today," Martin said. "We are staying at an airport hotel with our wives and children, we were in St Lucia and Trans Global offered us a flight here or to stay in our hotel in St Lucia and as we all have to get back to the UK as quickly as possible we opted for the flight here. There are about forty of us altogether and we do want to stay together if possible until we get back to the UK, most of us are now having problems with our credit cards and we don't have much cash left, we have paid for 2 nights at the hotel but we are soon about to run short of food, drinks and accommodation unless you can help us out," Mike said.

"Ok, obviously many UK residents who are over here on business or on holiday are in the same boat as you, so to speak," stated Martin. "We do not have unlimited funds here and we are also having problems getting any money transferred from the UK to here because of the financial meltdown that's also occurring due to the confusion caused by the eruption. We can give you some funds to keep you in the hotel until we can move you on, what's the name of your hotel, we will contact the hotel ourselves and sort that out for you. We can also help with subsistence allowances to give you some basic food, water and medicines, however anything more might be a little difficult currently." Martin continued, "getting you all on direct flights home to the UK is out of the question, even if there were flights going there, which there aren't, the poison gas cloud means that you couldn't get from any airport to your homes." "Even ships aren't going to Europe currently as the gas cloud is making life impossible for sailors to breathe." "All we might be able to offer is some sort of onwards flights to somewhere like Indonesia, India or Dubai or somewhere in that region, that is literally the nearest place we are getting people to currently, from there you would probably have to go overland or get a ship along the Med, however you won't get farther than Greece as the gas cloud is affecting parts of Italy, France and Spain now and depending on the wind directions is still slowly moving South although by the time it gets to Southern Europe and the Med the cloud is pretty diluted and has not caused as much damage as to Northern Europe. If you have women and children with you, we wouldn't advise taking them anywhere near the gas cloud, even half an hour exposure to this stuff is enough to bring on respiratory problems and with elderly and young people it's even more dangerous."

"We have heard yesterday that several European governments, including the UK, are negotiating deals with North African and Middle Eastern Countries like Egypt, Morocco, Tunisia, Israel and Lebanon to set up refugee camps where EU citizens will be housed, fed and looked after medically until this damm fog lifts. I hear that some of these camps will be pretty basic, old schools, civic centres or tent cities, it might have been better for all of you to have sat it out in St Lucia, although it's not my job to tell you what to do, I am just trying to advise you all," stated Martin.

Reggie then said, "it's all very well for you to say all that, however we all have family back home, brothers, sisters, elderly relatives, jobs, businesses and we can't just sit around here waiting who knows how long for this blasted eruption to slow down, perhaps we could take our wives and kids with us as far as the Middle East or India and leave a few of us there to look after them whilst the rest of us soldier on and try and find out what is happening back in the UK, we can't just sit here and do nothing."

"Don't you think that we don't have family, wives daughters, sons and parents as well who we haven't heard from for some time as well, we have to carry on working here sorting everyone's problems out and we also don't know for sure what's happening back home, it's extremely worrying for everyone," stated Martin, who was getting very frustrated that all these holiday makers didn't appreciate that they too had family and homes in the UK.

"Calm down everyone," said Mike, "look, we are all in the same boat, I'm sure you have had hundreds of travellers giving you grief, however we do need to get back to the UK, I am sure that this eruption will cease soon, just like the previous icelandic volcanic eruption did, with that volcano with the unpronounceable name Ejy.........kull, or something like that. If everyone else in our group agrees that the easiest route home would be via India or the Middle East we should try and work out where would be the best place to fly to, then perhaps we should leave the wives and kids there, then work out where we can get rail or bus routes across Europe and back to the UK."

"Well we had a group of businessmen in here yesterday," stated Martin, "and they had been faxing and phoning around and had worked out that they could get flights through to Dubai and then connecting flights to Cairo in Egypt, from there they would make their way by rail to Alexandria and from Alexandria they hoped to get a ferry through to Corinth in Greece and then take their chances with buses, trains and taxis."

"That sounds like the best idea that I have heard all day, we could leave our families in Cairo, and try and get this ferry over to Greece and take our chances from there on in, how long will it take you to get us sent over to Cairo Martin?" said Mike.

"Leave it to me gentlemen, I will get Sophia in our travel department to make the arrangements and then we will call your hotel to let you know when everything has been arranged, it might be a couple of days before we have any news for you, however I will also get Sophie to get you some money to keep body and soul together until its time to fly you all out, please go out of here turn right out the door, down the hallway take the second left into the travel office, take a number from the machine and Sophia will sort you all out."

"Many thanks Martin, you have been a great help," Mike said, they all shook Martins hand and set off towards the travel office prepared stoically to start queuing in another line of weary travellers.

Harold Webster peered out of his front bay windows looking towards the gravel drive leading down to the main road, Maitan, his young Filipino carer had not been at his house since Friday evening, now it was Tuesday morning 12th October and the poison fog had been swirling around his house since Friday evening, soon after Maitan had left him, Harold had watched the news and knew it would have been lethal for him, being elderly to have gone out in this fog, he had not seen any cars drive past since Saturday morning. The news wasn't good, the Government had not actually planned for this sort of contingency and he doubted whether there would be any social services visiting him soon. Harold, although he had a problem with Arthritis, could hobble about with a stick and although he was seventy three, he still had all his faculties and mental abilities, he had a fairly well stocked larder, cans of vegetables, packets of pasta, cereals and dried foods, he even had some camping gas from a camping trip that Laura had Mike had been on and had somehow together with all the other camping equipment ended up in his garage. What really worried Harold were news reports that as no one could get around due to the fog, there had been reports of intermittent power cuts and as the house was a large 1930's detached house on its own in the country he worried about power cuts and the temperature dropping and him being able to keep himself warm at nights, as well as cooking and eating which although were tedious he could manage that side of things, however the continuing reports on the radio were starting to worry Harold, as his son Mike was on the other side of the world and he had no way of contacting him.

Section 3

How to survive the upcoming Icelandic volcanic eruption

By Mark Reed

Harold listened again to the latest news bulletin and heard that many farm workers in the Peterborough area had actually been dropping dead in the fields around the Boston area. Exactly the same occurrence had happened in 1783, however at that period in time many thousands of farm workers had been dropping dead in the fields, as many as 21,000 may have died, the sulphurous fogs contained poisons that attacked the soft tissues in the lungs and respiratory areas. Many Poles, Latvians and Lithuanians, together with Rumanians and Ukrainians worked in the intensive arable farming areas in the Peterborough area of Cambridgeshire and Lincolnshire. Harold listened with growing trepidation and shuffled around the ground floor adding some more water to the wet towels put against all doors and draughty windows, as per the instructions given to all citizens by the BBC. The anchor continued, "The Government has called out the army and has provided many of the troops with bio haz suits and respirators, these units are now patrolling key government installations, power stations, TV transmitting stations, hospitals and supermarkets and some of the larger shops to stop any further looting. There has been unconfirmed reports of many supermarkets, food shops and off licences being looted by youths and other undesirable elements who have tried to take the law into their own hands. The Government has now made it abundantly clear that anyone caught from now onwards looting shops will be dealt with by the military authorities in an extremely harsh and quick manner, they will not be drawn on exactly what that terminology actually means."

"The PM has today again urged everyone to remain calm and to obey the military authorities, a curfew has now been imposed on all unauthorised movements between 8pm and 7am, only those members of the public who work in the following sectors are exempted, and proof of where you work will be requested by the authorities before you will be allowed to pass through any checkpoints. The exemptions are, power station workers, fuel delivery drivers, supermarket and food wholesalers, food and drink producers, farm deliveries and associated trades, hospital workers, government and local authority staff on the high priority list and other key utility workers regarding sewage plants, electrical and gas supplies and repairs and water supplies and repairs. These persons should have already received a respirator and goggles and instruction manuals in how to work in the current poisonous foggy conditions. Any of the above who has not received their respirator and goggles should ring the following number, which will be mentioned again at the end of this broadcast." The broadcast then went on to discuss further health and safety issues and discussions of exemptions from the curfew and other matters. Harold then rang Frederick Hawkins, one of his neighbours who lived about fifteen minutes walk away down the road.

"Hello Fred, can you hear me, its Harold from up the road."

"Dear lad, how are you coping, did your wonderful carer turn up and make your lunch today?"

"Sadly no old chap, and with all this swirling mustard gas, and what with my weak chest and wheezy cough I don't really want to start wondering around town today."

"Perhaps I should try and come over to your house and we can try and help each other to keep warm, cook and even just keep each other company, you know it's very lonely since Edith died for me to be rattling around here in this big old house, more of a mausoleum than a house," said Frederick.

"I am worried about you breathing all this muck into your lungs, it says on the news that hundreds of farm workers have died near Peterborough and they were mostly strong young men and women, not an old codger like you," replied Harold.

"Look, Fred, we put up with much worse during the Blitz, bombs going off all over the shop, if you think that a little bit of tinned London Smog is going to stop me getting around then you would be wrong. I will tie a small wet towel over my face, put on my old air force goggles and wear gloves and I will see you in twenty minutes," Quipped Fred.

Mike, Reggie, Vincent and Henry, together with attendant wives and children and the other families trudged wearily into the super heated air outside of Cairo Airport in Egypt, Mike was amazed by the heat which made buildings shimmer in the heat, they started looking for a minibus with driver for hire in the October afternoon (16th October), it had taken the group nearly 4 days of fighting through crowds of stranded tourists and travellers from Mexico City, Jakarta (Indonesia), Dubai and then finally Cairo, the airlines had temporarily agreed to pool their resources and were helping each other to try and get the hundreds of thousands of stranded travellers to their final destination.

The logistics of this operation by land, sea and air were enormous as fuel, spare airplane parts, crew and food and luggage were all in the wrong locations or not available, so airlines and shipping companies were making do as best they could. A sense of Camaraderie between all the different companies and organisations was helping oil the wheels of this huge undertaking rarely seen on such a scale since WWII and the Berlin Airlift of 1949. Airports had to erect marquees and tented accommodation to house the thousands of stranded tourists and their families. Bedding, blankets, food, water, medical supplies also had to be procured by the various authorities involved. Due to the dire conditions in Europe, European Russia and parts of North East USA and Canada, it was impossible for many of the travellers to actually get home, as home was under a dense thick poisonous fog which was continuing to roll across the plains of Europe and N E USA, therefore for the time being the UN had requested that the local authorities in each country should house, feed and clothe and supply emergency medical aid to all stranded travellers in each country until the emergency was over. The UN's main responsibility was to try to organise major food and water distribution across the affected territories and to ensure that basic amenities such as power, water, cooking gas was available to all citizens of the countries involved.

In reality, although the UN and EU and all the countries that had signed up to do all the good deeds that were listed in the central charter of the UN, in reality in a major pan continental disaster scenario, such as this one was turning out to be, many of the countries especially in the third world were not able or had the resources to comply with all the requests of the UN central offices in New York, for instance in Cairo where thousands of holiday makers were stranded, the authorities had attempted to put some of the tourists in small hotels in Cairo and the surrounding neighbourhoods, however the authorities barely could keep track of what was going on and the situation quickly deteriorated, there were no available tents for housing these stranded travellers and as more refugees started arriving from Europe things rapidly got out of hand and all attempt at controlling the situation collapsed.

Mike and his weary gang of travellers eventually managed to hire four minibuses at exorbitant cost and after the drivers had spent two frustrating hours crawling through the clogged Cairo streets, honking their horns every ten feet, they had all finally arrived at Cairo's main train station and after many further altercations and delays had finally taken a train to Alexandria in the Nile Delta.

The next morning Mike went to see the Andros Shipping Company near Alexandria's harbour area and finally went to see the shipping clerk, Mehdi, who was organising passenger movements on the front desk on the ground floor of Andros's offices.

"Good morning," Mike said, "do you speak and understand English?" "Yes sir, we deal with all Europeans here on a regular basis, how can I help you?" Mehdi enquired.

"I need to book passage for about ten families of English and some Dutch and Germans, we all want to get home to our countries as quickly as possible, there are about five English families, about twenty five persons who want to get back to the UK as quickly as possible can you help us?" Mike intoned.

"Unfortunately all British ports are closed and hardly any trains are running from Southern Europe to North Europe, the forecasts are that the eruptions might last several months or more. All I can offer you currently are spaces on a mixed car, freight and passenger ferry going to Istanbul in three days time, which will cost each of you $ 500.00 each, such is the demand for these tickets that some unscrupulous touts are demanding triple this figure from other travellers and they are getting it too." Mehdi stated.

"This is outrageous, on your own booking forms it states the price to Istanbul as $ 150 per person, how come it is now $ 500 per person," Mike said in a heated voice.

"Dear sir, I am but a humble shipping clerk, I do as I am told, in these troubled times, unfortunately my boss's business is also in much trouble, many of our routes are now closed to business because of the fog of death, who knows when it will clear up or our routes will become open again, we have to try and recover some of our losses, its supply and demand basically," Mehdi stated, with a flourish of arm movements with both palms extended outwards as if implying it was the will of god and there was nothing he could do about it.

"It sounds ominously like rip off any poor foreigner who is stuck abroad and get what you can off them. Look, anyway I have to consult with my fellow travellers and then I will return and we can sort this out, have you many places left?" said Mike in a dejected tone.

"I can reserve your spaces for a down payment of $ 1000.00 cash, there are some spaces left, however they are selling fast," said Mehdi.

"I will speak to my friends and get back here as soon as I can," with that Mike strutted out of the shipping company, fuming at the situation that was causing problem after problem with his intentions to return to the UK.

Mike then went back to meet with the others and after much wrangling and discussions it was agreed to leave the wives and children in Alexandria with Henry, due to his heart condition and his age, they would pay the local hotelier enough money to keep the wives and children looked after for about a month and then the younger members of the group would attempt to go from Istanbul across Europe by car, train or hitch hiking to see how their elderly relatives and other family members were faring in the UK.

Harold peered out of his lounge windows for the umpteenth time to see if he could see Fred coming up his gravel path, all he could see was gusts of billowing yellowish fog banks. He was just going back into the kitchen again when he saw someone come stumbling up the path staggering right and left, he noted that it must be Fred so Harold hobbled as quick as he could to the front door to let Fred in.

Fred staggered into the hallway, he had a damp towel wrapped around his nose, mouth and lower face tied around his neck, he wore an ancient pair of flying goggles and a flat tweed cap on his head, he pulled down the towel and gasped for air.

"What a bloody nightmare, it's unbreathable....," he started a coughing fit and Harold gently led him into the lounge and sat him down near the fire, he was wheezing badly, his lungs after 60 years of smoking weren't in a good condition to begin with.

"Fred, take it easy, relax, get your breath, let me get you at tot of whisky, don't say anything for a few minutes." Harold hobbled over to the drinks cabinet and poured Fred a double of his most expensive Talisker single malt whisky. Fred had a few sips still wheezing horribly, Harold was really starting to get alarmed at the state of his old friend and neighbour.

"Forget pea souper," Fred wheezed, "that stuff seems to have come straight from Ypres and the mustard gas attacks the Jerries sent across in 1915 and 1916, not that I actually was there, before my time," Fred expounded. "The stuff really stings the lungs, you can't breathe any air or oxygen out of it, it sort of clamps itself around your insides. Luckily for me it wasn't very thick outside my house which is higher up on a hill, as you know, it was only when I was approaching your house that I got a lungful of that poison, if I had to go on another few minutes I would have been a goner, I pity anyone who has to actually work or go out in that stuff, without breathing gear no one will last longer than a few minutes and it stinks of rotten eggs." Fred's wheezing calmed down after about half an hour of resting, however there was a bad rattling sound to his breathing that Harold was worried about and he was thinking of calling Dr Smithers from the surgery down at the bottom of the hillside.

"Fred, I'm going to call Dr Smither's, that chest of yours sounds like bronchitis and it doesn't sound right, does it hurt you when you breathe?" "Harold, at my age it's easier to tell you what doesn't hurt, it would be a shorter list, I do feel wheezy, however I always feel wheezy, I have cut the fags down to two a day, you will be saying no Port, no Whisky, no G & T's next."

Harold then decided to take matters into his own hands and rang Dr Smither's, a recorded message informed Harold that the surgery was closed due to the poison fog emergency and to ring the special, "fog alert NHS emergency hotline." Harold rang the number and was asked various questions by a machine, eventually an operator came on the line, "fog emergency NHS hotline, Amanda speaking, how can I help you," "my good friend Fred has just had to walk to my house in the fog and he doesn't sound too good, his chest has a bad rattling, wheezing sound when he breathes can you send a paramedic crew out here to see to him tonight." Amanda replied, "my job is to discuss the situation and then prioritise the action necessary, we are being inundated with calls currently, we have to prioritise patients for triage action to take place, is your friend conscious, can he speak." Harold replied, "yes, he is conscious, he is wheezing badly, he was a smoker, he is over 80 years old and I think he is in pain right now." "Keep him warm," "Amanda replied, give him water to drink, keep him upright so any fluids in his lungs won't affect him," Amanda continued "what about his medical history, has he had pneumonia, bronchitis or any other respiratory illnesses within the last twelve months," this Q & A session continued for some minutes, then Amanda took Harold's address. "We cannot state when help can be sent to you in your area currently, all local hospitals are full, our services are being incredibly stretched currently and the Royal Army Medical Corps are also helping as is St John's Ambulance Services and other voluntary groups, all we can advise is that your details will be sent for triage assessment by the rapid response team over in Gerrards Cross which is your nearest Medevac station. Please keep your friend awake and give him some hot drinks and keep him warm, we cannot give you a time that someone will come and visit you yet. If the phones don't go down again ring us in 3 hours time, here is your reference number, 6348765GRT, good luck." "Good luck to you too," Harold muttered under his breath.

"Well old chap," wheezed Fred, "is the ambulance on the way with some dishy nursey to give me a bit of mouth to mouth and sort out my old battered chest," "not exactly old boy, it appears that the NHS is inundated with calls, everyone is out and about, hospitals are full to bursting point and they can't give me a time when someone will be able to get out here, looks like you and me will have to fend for ourselves a bit longer, don't worry, we will see this wretched fog out, keep smiling, I am going to make you a nice strong cup of tea and get you a blanket to wrap you up in old chap."

Mike, Reggie and Vincent peered out of their cabin window, the captain had ordered all windows, doors and hatchways closed several hours earlier as the cargo passenger ship was making its way from Alexandria to Istanbul and reports from the weather centre in Ankara in Turkey had informed the captain that some parts of the Mediterranean had fog banks rolling across the sea, although being this far south the fog banks were weaker than Northern Europe and only in scattered locations, however Mike's blood ran cold as he saw the swirling yellowish fog close in around the ship and apart from the reassuring sound of the ships diesel engines reverberating through the hull there was a sort of eerie silence surrounding the banks of fog. "Anyone fancy a run around the deck," Reggie chirped in, "after you old bean," said Mike, trying to lighten the mood a little, "we should have stayed in St Lucia, lobster, steaks, surf and sun," said Mike, "we all had the option to stay and we all agreed to try and get back home," Vincent reminded them. "If it's bloody like this here then what the hell is it going to be like back home, I daren't even think about it, I'm so glad we left the wives and kids back in Alexandria," said Mike.

Fred had dozed off in Harold's favourite comfy armchair, Harold was very worried about the rattling wheezy breathing of his close friend. All these years of cigars and cigarettes have weakened his lungs and even though he had cut down recently, this damned fog could have made him very ill, thought Harold. Harold was also worried about what was happening to his son Mike and daughter in law Laura and the two kids, Harry and Lucy, on holiday in St

Lucia. Harold hadn't heard from any of them in nearly a week and he was extremely upset as Mike usually called him to say hello every evening or at the least every other day and his son was his main lifeline to the world. He did have another son David who lived in Chicago and was married to Lizette an American girl that David had met whilst on student exchange at Syracuse University near Niagara Falls. The phone had gone dead again, this was the third time in a week. Harold didn't know it but behind the scenes the Government had relocated to a secret underground facility near Northwood in Middlesex where one of the cold war bunker cities had been refurbished in the late 1990's due to a new perceived threat from terrorists with nuclear weapons. After 9/11 work had been progressed very quickly so now Stuart Rumbold the Prime Minister held court with his various ministers, generals, admirals and technical support staff and army personnel in an underground city of dormitories, store rooms, communication centres, generator rooms, and a command room which had a large glass style map of the UK covering one of the walls with all major urban centres picked out with orange lights and motorways as green illuminated lines and railways as blue lines. On a big central table set around with chairs were other maps, sheaf's of papers, laptops and mobile phones and BlackBerries.

"What's happening to the land lines Edward, please explain that again," asked the PM, "well prime minister," Edward Grovemont who was permanent under secretary for the minister in charge of telecoms and communications started to speak, "It's the same story, we only have a limited number of these breathing respirators for the troops who are out there searching for the key personnel who are needed to repair the telecoms mainframes, when bugs get into them and they also break down, these rolling power cuts aren't helping as the telecoms control centre's are being blacked out occasionally and then the whole system trips off and the backup systems get overloaded and the whole thing grinds to a halt, we have to contain the black outs and put resources into getting secure power to these communication centres."

"Rubbish," snorted Blake Edmondson, NHS Minister, "we need to ensure that all hospitals get priority with regards to power supplies, that's the only thing we should be concentrating on right now, every time the power goes off, half a dozen hospitals around London alone have to go onto emergency backup generators and we only have a limited amount of fuel at most hospitals to keep these generators going for a few days, if we don't get re supply and there are more power cuts then that's it, no operating theatres, no heating for hospital wards, no nothing." "Hang on a minute," said Jock Boydson, Minister for Transport and Distribution, "priority must go to warehousing, ports and dock facilities and getting the army to locate where all the HGV drivers are stuck, without distribution of oil, food and heating fuel, we are all finished."

"Hang on everyone," said Stuart Rumbold the PM, "we need a co ordinated approach to this crisis, all we seem to do is go round in a big circle, everyone quite rightly fighting for their share of whatever meagre resources we have at our disposal, however there are millions of frightened, lonely, cold people out there who are depending on us to make the right decisions and the correct judgement calls as to how we get though this mess and ensure that we do look after our citizens, they have a right to that, now stop this endless bickering, lets once again try and formulate a plan to deal with the most pressing crisis, which I see as, firstly, we need to work out how to get enough staff to power stations to keep them going. Secondly, we need to direct some secure power lines to at least one of the phone networks to keep it on the air, thirdly we need more troops in Land Rovers with respirators and goggles to try and get the HGV drivers with all their breathing gear into the cabs so we can keep the oil and food rolling from the ports."

General Cranbrook, Army Chief of Staff butted in, "Prime Minister, these respirators that we are using are not holding out too well, they must be 25 to 30 years old and were part of our nuclear and gas attack preparations, troops cannot wear these contraptions out there in this poison fog ad infinitum, I would say that they should only wear them for three hours at a stretch and then breathe ordinary air for at least two hours, otherwise the troops will all start getting respiratory illnesses as well as the public." He continued, "some of our boys could drive some of the HGV's if they had to, as they have been trained on Army HGV's so they know he basics, I could put a call through to GCHQ and work out how many of our boys we could get into the lorries to keep things moving."

Francis Moffat, the senior chief of the Metropolitan Police, who temporarily was in overall charge of all police forces across the country said, "lorries have to be accompanied by at least 2 Land Rovers with a full complement of armed soldiers or police, we have been getting several distressing reports of lorry loads of Tesco's food lorries being hijacked up near Sandy on the A1 North of London, the drivers were shot and the vehicles have now disappeared, this is probably the work of local vigilantes who have decided to take the law into their own hands as they cannot see any help coming from us, the Central Government so they are robbing from the supermarkets to feed their local populations." Francis continued, "they seem quite organised, they even had respirators, there must have been some private stockpiles of respirators that have fallen into this crowds hands." Francis continued, "I have got all available men in the area looking for these characters, and they are not the only examples of this lorry jacking epidemic, up near Liverpool, three lorries disappeared yesterday and over near Folkestone, another lorry has gone, we must have better protection or some sort of convoy system with plenty of armed police and soldiers riding shotgun for these lorries to be properly protected."

"What's the petrol and heating oil situation like Stanford," asked the PM. Stanford Crowethorne, Minister for Energy, peered through his bifocals at his notes, irritably waving away one of his junior staff. "We have a strategic final reserve for the armed forces, police and ambulance and fire brigade of about two weeks left," he intoned, reading from his notes, "unfortunately the commercial sector has all but run out of fuel and we are concentrating on organising convoy systems from Milford Haven refinery in South Wales to various strategic power stations in the Midlands around Manchester, in the London Area, Birmingham, Newcastle and a few other cities. However, demand is too great and some power stations have already run out of oil, so we are having to start rolling power cuts across the country and try and keep as many hospitals, old age homes, mental institutions and army and government organisations on line, however as ships are not able to dock at Milford Haven due to the poison fog clouds the situation, which I admit is pretty grim now is set to get worse in a few days."

"Things are not looking too hot for us over here? is there any chance of help from France or Germany, we pay millions of pounds every year to stay in the European Union, let's try and get something back for once, apart from more red tape," said the PM, there were a few smiles around the table at these words. James Hawthorne, the Foreign Minister, cleared his throat, "Hmmmph, I have been on the phone to Monsieur Montand the French President and have requested that gas supplies be re started again, he informs me that as no staff further East are manning the gas switching stations in the Czech Republic and in Belarus that they cannot obtain any gas themselves, they have protested to Moscow and Moscow is doing what it can to restart supplies, however nothing concrete has appeared yet," James continued, "The French appear to be suffering the same as we are except, they had the hindsight twenty years ago to build a new generation of nuclear power stations across the country so whilst we freeze our assets off in this murky fog, they are as warm as toast with all lights around France blazing merrily away and heating and phones and power for everyone.

I have asked for some power to be diverted through the undersea cables across to keep London going and he is considering the request and will let us know tomorrow." "Bloody Hell," fumed the PM, "They still haven't forgiven us for Waterloo and Mers El Kabir" (where the British Fleet destroyed the French Fleet in 1940 to stop it being turned over to the German Navy by the Vichy French Government), "If the bloody tables were turned we wouldn't stand there and say, well as Duke William of Normandy killed King Harold at the Battle of Hastings in 1066 we are not going to give you any power !!!!" "This is hopeless, what about the Dutch and the .Germans?" James replied, "The Dutch don't have a direct link into our National Grid, it would have to go through France and anyway they are running out of oil and gas too. The Germans would like to help, however they have been put upon by the Danes and the Belgians and are trying to help them and they also would have to send any power through the French Electric Grid, I am sure we will work something out, however it will take a few more days in my opinion PM." "We don't have a few more days, elderly and nursing homes without heat, water or power will mean people will start dying today unless we can sort something out, what about our remaining Nuclear Power Stations are they still operating?" Stanford Crowethorne replied, "yes PM, the few Nuclear Power stations that we have left in Winscale, Dungeness and a few others around the country are hooked up to keep most of the hospitals, some nursing homes, GCHQ in Cheltenham and some of the Government Ministries going, however they cannot do much more than that and they are mostly off their peak performance now and are winding down soon for de commissioning, as you know the next generation of nuclear power stations are not even off the drawing board and won't be up and running for ten to fifteen years at the earliest."

Harold had been dozing for several hours as had Fred who was wheezing and rattling in a very disturbing way and Harold was getting more and more worried about his old friend. At about ten pm Harold heard the sound of wheels on his gravel drive and saw faintly defined headlights piercing through the swirling yellow fog, an army auxiliary Land Rover squealed to a halt in front of Harold's house with a scattering of gravel stones against the front steps. Harold peered through the murk and saw an army soldier in a khaki green boiler suit with a gas mask on, get out of the driver's door and a paramedic dressed in a strange white suit with red cross armbands and a funny looking strange gas mask on get out of the passengers side, the paramedic was carrying what looked like a large briefcase. They crunched their way across the gravel to the front door and rang the bell, Harold scurried off to let them in, he held a handkerchief to his mouth and nose as he opened the door and the two boiler suited men quickly came into the house, Harold smelt the foul smelling poison fog smelling of rotten eggs and pricking the back of his throat as he slammed the door back again and pushed the wet towel quickly back against the bottom of the door.

The soldier and the paramedic removed their military looking gas masks and breathed deeply of the clean warm air in the house. "Good evening gents," said the soldier, "I am Corporal Miller and this is Medical Officer Harvey and you two are extremely lucky that we are here, you must be Frederick Hambledon," said Corporal Miller, pointing at Fred, who was having a coughing fit and trying to nod at the same time, "Oh, we are so happy to see you both," said Harold, "Fred and myself had practically given up on ever seeing another living thing again, I don't know how anyone is getting around in that foul pea soup, I haven't seen or heard anything on the road for a couple of days now," wheezed Harold who was still struggling to gain proper breaths after his encounter with the fog at the front door. Royal Army Medical Officer Harvey knelt down and started to examine Fred, using his stethoscope and listening to Fred's chest, as Fred stopped coughing but continued wheezing.

"We were up at the Quaker Tea Rooms where the manageress who is pregnant has been stying and she has been having chest pains so we were seeing her, that was our 12th callout of the day so far and we were just about to go back to base at Gerrards Cross when the radio called and it appears that someone called Dr Smither's contacted our C O and asked him to pop into to see old Fred here, it appears old Fred is an old RAF pilot, so we felt we would pop over and have a look at him before turning in," said Corporal Miller. "Thank God you came by," said Harold, "let me put the kettle on and get you two lads some tea and a biscuit, what on earth is happening out there? how long is this dreadful stuff going to be here? and what's it all about?" quizzed Harold.

"Well, as far as we know, some old volcano in Iceland called Laki has erupted and instead of ash, or as well as ash it is spewing out some sort of poison gases which are blowing right across the UK and over into Europe as well we are told, everything has come to a standstill here in the UK, only people with proper respirators like ours can move around, the stuffs lethal even to young people, some kids over towards Uxbridge were trying to loot a Curry's Electrical Shop and although they had scarves tied around their heads they collapsed in the fog, by the time anyone could get there they were dead. There are also plenty of homeless people and other travellers caught outside that have been found dead locally, the Government has put out a bulletin that only the military and emergency personnel are allowed out and only those with respirators, we have been on the go for about 15 hours now and we need some kip very soon," said Corporal Miller. "What's the prognosis doc, will I survive to see the next Biggin Hill Air Show?" wheezed Fred, who was now feeling a little better and had stopped coughing.

"You should know better sir, running over here in the middle of all this poison gas, don't you listen to any of the regular broadcasts that the Government are making about no one to leave their homes and everyone to wait until help gets to them?" said Medical Officer Harvey. "I couldn't leave old Harold all alone, how would he cope, he can hardly walk and just about gets upstairs and his helper hasn't turned up, us old soldiers have to stick together, you don't desert a comrade in time of need," retorted Fred. "Well I would have to say sir, that you have been extremely lucky, at your age and also a smoker, running around in this muck, I think if you keep warm and stay inside and take it easy, within a day or two you should be Ok, there is no point us taking you back to base, as the Chalfont's and Gerrards Cross Hospital is overflowing with respiratory cases of people who got caught in the fog and severely damaged their lungs and I think it would be worse taking you there than leaving you here with a couple of inhalers and a nebulizer, said Officer Harvey."

Harold returned with the tea and biscuits and they all drank their tea and listened to the 11pm news broadcast on the BBC. "This is the BBC 11pm news broadcast, the main news is that a huge poison gas cloud from the major Icelandic eruption of the volcano know as Laki is continuing. Most of South Eastern Iceland has now been completely evacuated, however it is estimated that up to 30,000 cattle and sheep have died from poison gases so far as well as an unknown number of horses and wild animals and birds. In the UK, From the Scottish Isles to Cornwall, the poison gas has enveloped the country and many hundreds of deaths have so far been reported across the county, as well as many thousands of cattle and sheep and pigs and chickens have also died. The Government has again urged everyone to stay indoors and to put wet towels over any cracks at the bottom of doors and over draughty window frames to keep the poison fog out of your houses and flats. The Government is attempting to organise food distribution for when the fog lifts, although scientist predict that the fog could stay over the UK for several weeks, as there is currently a ridge of high pressure over the UK which is only expected to slowly move East over a two week period once the normal winds get through to the UK they should be able to blow the poison fog away to the South East. Some looting has been reported and the Government has ordered troops in respirators to patrol all cities and towns with orders to use whatever force is necessary to restrain looters, a curfew has been imposed between 7pm and 7 am every night, anyone who is not in the emergency or armed services caught outside after this time will be arrested and incarcerated immediately.

Many container ships and oil tankers have been anchored around the Canary Islands and the Azores, as their captains will not sail them into waters, where poison gas clouds have been seen rolling across the seas near to the UK and over the English Channel. All travel across the channel and under the channel through Eurotunnel and the Eurostar train service has been suspended for the duration of the poison fog. All air traffic has ceased between the USA and Europe, we understand some flights from Miami International Airport to Casablanca are getting through, however there is currently no scheduled services and many aircrew are trapped in the UK and Europe in this poison fog. Most of Europe from Norway down to Spain, Italy and most of Greece has also been affected although we understand that in parts of Romania, Bulgaria and Serbia and Belarus the fog is thin enough that some rail and road traffic is moving, however it is intermittent and no one can tell if things are going to get better or worse currently." The anchor continued "Stuart Rumbold, the Prime Minister has asked that everyone stay calm and try to carry on with their daily lives in a normal as fashion as possible, the PM will make a keynote emergency speech about this crisis tomorrow at 2pm. He has asked The Army, RAF and The Royal Navy to ensure that the power stations, water works, gas supplies and sewage facilities are given priority in order to keep everyone safe at home throughout this emergency, we have heard of rolling power cuts across the country as key members of staff who run the power stations have not been available or could not be found, Ministers are doing all they can to ensure continuity of supply, however with no oil tankers reaching the terminal refineries, the Government is in a very difficult position and difficult decisions will soon have to be made as to where to put resources, hospitals, old people's homes, schools and other civic centres are believed to be top of this list of priorities.

Scientists are now working to see what effects the fog is having on reservoirs and drinking water, as it is feared that the longer the poison gas cloud interacts with the water supply the more chance there is of supplies becoming contaminated and the water becoming too acidic to drink, scientists are now conducting test around the London area and hope to release their findings over the next two days," the anchor continued speaking, Harold then said, "we have some food and bottles of water, however if they cut the gas off how are two old codgers like us going to survive, its getting pretty cold out there at night, it's down to about just 2 degrees some nights and soon it will be below freezing?" "look, we will probably be coming this way tomorrow to see the pregnant lady up at the Quaker Tea Rooms again, so we will try and get a couple of paraffin heaters or oil fired stoves, which I think I saw in the emergency store room back at HQ," said Corporal Miller, "thank you lads," croaked Fred, "just seeing you two has put some life back in me, it's the feeling of isolation that gets me, that's why I tried to get over here to Harold's house, I couldn't bear just sitting there, all alone."

"Don't you two old troopers worry, we will be back tomorrow at some point with some grub, maybe a bottle of whisky and some sort of heater," said Corporal Miller, "don't come near the front door, you two stay put here in the nice warm lounge, we will suit up and let ourselves out and then after a few minutes come out and push those wet towels back against the door frames, good luck, keep warm and out of that damned fog, hopefully we will see you again tomorrow," said the Corporal, then, they suited up and re attached their respirators and left the house and then slowly drove away.

The passenger carrying cargo ship finally docked at Istanbul in Turkey and Mike, Vincent and Reggie and the others disembarked onto the bustling quay side, all was chaos, crowds of locals trying to do business, selling trinkets and offering various services, tourists and travellers sitting in dejected groups arguing with mini bus drivers and ships officers, the heat wasn't too bad as in October in Istanbul the temperatures are down to the 70's and early 80's, not as hot as Alexandria and Cairo.

The small group eventually managed to negotiate an extortionate rate with a local mini bus driver to take them to the British Consul in Istanbul and feeling tired, and dejected they sat as the minibus driver honked and swore his way through the congested streets surrounding the Great Bazaar and through the centre of the city to the British Consuls office, the thought of going through the same rigmarole as Mexico City didn't appeal to Mike, however they had to find out what options were open to them to try and get themselves back to the UK.

Eventually they arrived at the British Consuls offices and although there were a few people wandering around the

offices, it wasn't as hectic as The Mexico City Embassy
had been. After waiting for about an hour they were led
into a first floor office to see Mr Foster Humboldt CBE,
Her Majesty's Honoury Consul to Istanbul. "Apologies
for the wait, I have been up to my neck in trying to make
arrangements for many stranded and worried Brits over
the last week, how can I help you gentlemen," said
Foster. Mike laboriously explained their predicament
and their travels since leaving St Lucia and their worry at
not being able to contact their relatives in the UK, after
he had finished Foster spoke, "sounds like you chaps
have had quite an adventure, you are not the first group
of travellers with that story or a similar story to that one,
that I have heard over the last week. I will give it to you
all as straight as I can, its no use faffing about with
possibilities and story's of what we can and cannot do
here. I am afraid you have come to the end of the line
gentlemen, about 3 hours rail travel across the frontier
and into Bulgaria all rail, road and river travel comes to
an end as banks of yellowish looking poison fog have got
past Sofia now and are quite close to the border with
Turkey in some places. You can't move about in the fog
unless you have a respirator and goggles or the stuff will
kill you. Although it is patchy and does not cover all of
Bulgaria, there are some areas it hasn't covered as it is
still moving around no one will chance it apart from
military personnel and they will only move with full Bio
Haz suits and respirator equipment. I know what your
next question will be and the answer is no, a Bio Hazard
Suit and respirator kit will cost you about $ 1500 here in
Istanbul and the price is going up daily." Mike said, "we
have to try and get back to the UK, do you know what
conditions are like now in the UK?"

Foster replied, "you must have seen the news headlines and heard the BBC World Service reports, things are pretty grim in the UK right now, oil and petrol have just about run out, supermarkets and shops have been emptied and looted, lorries with food are being hijacked and oil tankers can't get into ports with their oil due to the fog, people are battened down and have been told to cover all draughty areas with wet towels to keep out the fog from their homes, the phone lines and mobile lines are all down and things aren't looking good. The Icelandic Government has had to evacuate Reykjavik and has relocated over to Montréal in Canada, the volcano Laki is still erupting, belching out this vile gas, however some of the instruments around the eruption have shown a slight lessening of the volume of gases being ejected currently.

The other problem is the wind direction, recently this high pressure over Europe and the UK has meant that the fog slowly rolls across Europe and more or less stays where it has gone as no wind has blown it away, however the Met Office has predicted that south easterly winds could pick up next week and some low pressure over the Azores could push into the high pressure towards Iceland and change the weather patterns and start to blow this fog away, we can only hope that The Met Office get this one right. Mike replied, "we have been following the news stories, has no one even attempted to try and get back to the UK? Is there any chance that we could go back with any Army or RAF personnel that might be based around here or perhaps Cyprus?"

The main reason for me inserting these narratives into my book is to show vividly how life could come to a temporary stop as we know it, during some sort of serious Laki Style eruption. No governments across Europe as far as I am aware have made any preparations for such an eruption. The Icelandic Prime Minister himself in a broadcast soon after EJ started to erupt in March 2010 warned that in his estimation, this was the beginning of a bigger eruption cycle by Katla or another of these eighteen extremely volatile volcanoes sitting on top of a hot spot in the Mid Atlantic Rift. The human impact on an individual scale will be a life changing event, if the power goes out on a regular basis and there is no food in the shops and gas and heating goes off in winter, suddenly from being online with our laptops and BlackBerry's and i phones, we are plunged back into the Middle Ages, many of us will not be able to survive these cataclysmic changes unless you have really prepared for these events yourself well before they happen.

It does not have to be a Laki style eruption to cause major mayhem, ordinary Icelandic strato volcanoes and ordinary volcanoes are able, if the wind is blowing to the South East from Iceland, to cause much disruption to life on the European Mainland by stopping all air traffic for months on end and the ash fall and acidic rain and possible sulphur fogs and other atmospheric disturbances including extremely cold winters, would all ensure that crop production and chicken, sheep and cattle rearing may be severely disrupted for some time, therefore although I do not think one of these strato volcanoes will kill many thousands of people from direct effects, e.g. lava, lava bombs, spheroids (unless from a super volcano), apart from the Laki style killer poison fogs, what will severely disable the population is the fact that only about 5 % of the UK population is actually involved in any sort of farming, the rest of us are involved with service industries, industrial manufacturing and government and local authority employees. Therefore, we will be totally reliant on the Government to help with the distribution of drinking water and foodstuffs, heating materials, blankets, emergency repairs to houses and flats and similar scenarios.

I don't think there are any Government plans in the UK or the rest of Europe which has for a moment thought about all these contingencies, as per my short earlier narrative the elderly and infirm will probably be left to the mercy of neighbours and family and friends and those trapped abroad as we saw from the recent light dusting of ash cloud activity, when some people were trapped for up to six weeks abroad, if a real situation emerges it may well be totally impossible for hundreds of thousands of citizens to return to their home country for many months, and as my little docu drama shows, who is going to pay to feed, clothe and house all the hundreds of thousands people trapped abroad?

The British Government, if the eruption started in say mid August may well have about three to five million Brits abroad, as far away as Thailand, Australia and South Africa, South America, the USA and other far flung resorts, all trapped and at the mercy of local hoteliers, who might help out for a few weeks at most, however even with the UN involvement, someone is going to have to co ordinate the paying of all these local hoteliers and guest houses across the globe (as are other nations going to have the same problem), with the present massive cuts in public expenditure and the problems caused by the financial meltdown of 2007 and 2008, I doubt whether the UK's Consular Service abroad are geared up or even briefed on the multifarious possibilities of problems that they are going to encompass should the worst come to pass, which I believe it will during the next 5 years or so.

So, if you are abroad and a serious Laki style eruption occurs, you will probably be better off staying where you are and harassing the UK Consular Services to settle all the hotel bills whilst the emergency is ongoing. If you are abroad with your wife and children, then that is the safest place to be. Even if you did make heroic struggles to return to the UK to look after elderly or infirm or other relatives, it will, unless you have heeded this book and made all necessary precautions be extremely difficult to get petrol, food and heating materials. However of course I would if possible, in any event as long as I would know that my wife and children will be cared for at the resort, try and get home if you do have an elderly parent or relative who are on their own, they will need to be cared for. Obviously if you have other family members in the UK and you can contact them to ask them to look after your elderly or infirm relatives and they can help, that would take a great load off my mind.

I would hope that the first section of the community that the Government helps out will be old people's homes, mental institutions, hospitals and those in society who cannot help themselves. However, I have lived in the UK all my life and I know full well that quick reactions to events, forward strategic planning for these sorts of disasters are not really top in the minds of those who govern us. The French have had TGV high speed trains for about 25 years or more, we are still relying on trains built about 30 – 40 years ago and no replacements have even been laid down yet, average speeds on UK track are so slow that our French and EU partners must laugh at us. The slightest fall of leaves, the wrong sort of snow, a cold snap and the country grinds to a halt, can you imagine what it would be like, god forbid if there was a real emergency to handle, it does not bear thinking about.

My only consolation is when I think how unprepared the UK was in 1939 to face an implacable enemy, the German War Machine, we had left all of our weapons on the beaches of Dunkirk and we were on our own, the Germans had thousands of aircraft, thousands of tanks, heavy guns and trained troops, however once the nation got behind Winston Churchill and the Coalition Government and we all pulled together, and although it took six long years and the help of the Americans in the end the tough, resilient British Spirit of not giving up and seeing the job through defeated our enemy. I am confident although the new enemy is a natural phenomenon and not an invading army we will, after many hardships and many casualties and difficulties hang on and survive the poison fogs, molten hot spheroids, ash clouds and bitter cold winters and pull through, after a few years in the same spirit as our fathers and grandfathers had done, during the last war. I really do think the eruption will be as bad for the UK as the effects of the last war, however it won't be bombing raids over cities or U Boats sinking merchant ships it will be an unpronounceable Icelandic Volcano that will be wreaking havoc with the UK and Europe and our old foes will and are now our staunch allies such as, the French & German Nations in the upcoming battle to survive the wrath of the elements.

There is not much that you can do if you are away on holiday with the family or alone on business or pleasure, you obviously cannot carry 300 crates of canned food, bottled water and dried food away with you on holiday, the staff at Ryanair would have a fit and can you imagine how happy Mr O 'Leary of Ryanair would be with all the extra revenue from all the heavy suitcases, no we can't have that, can we? However, there are some things we can do to cover our health and well being whilst we are away.

1. Medicines and First Aid Equipment

We should, if we are on any sort of medication, trot off to our GP and tell him we are planning a three month round the world trip and can he give you enough of your prescriptions to last you three months. Especially if you have any serious illness, heart complaint, diabetes, MS, HIV, or if you are on a course of chemotherapy, mental illness or any other serious condition, the last thing you will want is to be trapped in St Lucia where they may or may not have any of the pills and medicines that you need and even if they do have them and you do have travel insurance they may still require you to purchase them and then reclaim the monies from your travel insurers once you get back to the UK. If the eruption goes on for months, things might get very difficult, as some of these small islands, like the Seychelles, Mauritius, St Lucia may then not receive their regular supply ships or planes, due to the Volcano's disruption elsewhere in the world. If St Lucia's weekly containers come by ship from Rotterdam, a major container port, and no ships are leaving or entering Rotterdam due to the poison fog or ash cloud activity then everyone on the Island will start feeling the pinch and if you require specialist gluten free foods or nut free foods or similar, these complicated diets and medicines will be the first things that run out on these Islands. Incidentally Ibiza, Majorca, Tenerife, Lanzarotte and Grand Canary are also Islands, as is Cyprus, so these things must be thought about prior to departure.

Other items to take might include a small first aid kit or some clean sterile hypodermics (sealed), Savlon Cream, Anti Histamine, Asthma Puffers, and spare refills, Sulpodene, Panadol, Aspirin, and all the usual ointments and lotions used abroad. You will have to use your common sense, there is much more possibility of getting medicines and help on Sardinia than Tristan Da Cuna or St Helena, so common sense must prevail here. Currently UK Customs & Excise Regulations and many other countries regulations ban the taking of more than 100 ml of any liquids through customs in hand luggage, however I think you are allowed to put these medicines in your main luggage which is put in the ships or aircrafts hold.

Without going overboard on this subject you could also take some water treatment pills and maybe if you are going somewhere remote like The Pitcairn Island or the Galapagos Islands, there are some spaceman style foods that re hydrate and self cook and can give high sugar, carbs and protein in an emergency situation. Again you will have to decide what is a priority and use your common sense here. If you are going on holiday to an area which you know has communicable diseases, don't forget to go on a six week course of anti malarial pills firstly and take some spare pills to top yourself up in case you get trapped out there. If the foreign country you are going to has west nile fever, cholera, typhoid, dengue fever or similar, I would suggest trying to get your hands on some anti biotics, if you can persuade your doctor to help you, all the better.

2. Finances

Currently if, when I go abroad and use my Visa Card, I always get a nasty surprise when I return to the UK, the rate that the credit card companies seem to give is always much worse than rates which you can buy yourself before going abroad. Currently you can actually get good deals at M & S and the Post Office when buying foreign currencies. I would in light of this book always try and take with me at least twice the normal amount of money you would normally take abroad, as you can see from my earlier "little story" when the chips are down and you need to get a ferry or hire a car in an emergency and the phone lines are down so your Visa Card does not work, hard cash is always going to win you through. That is why during the last war aircrew and soldiers had emergency packs including gold and silver coins to be used in an emergency, especially in the Western Deserts of the Sahara and other parts of the Middle East.

Although I am an avid fan of the Pound Sterling, I have unfortunately found on my travels that the American Dollar seems to be the widest accepted form of currency worldwide. Again if you are going to Le Touquet for a weekend, I am not advocating re mortgaging your house and going to Goldman Sacks to purchase thousands of dollars, common sense must prevail here. It is probably worth getting some local currency and some dollars, I doubt whether travellers cheques or credit cards will be of much use if a serious eruption takes off, central computer terminals may lay unmanned as staff will be unable to get to work, power cuts and logistic problems may affect Visa and other credit card computer terminals, so again cash and gold will be king in these types of emergencies.

3. Communications with home

These days when I travel abroad I do notice that most people are now carrying laptops, BlackBerry's, mobile's and other communication devices with them. The new EU Directive forcing the mobile phone companies to drastically reduce their "roaming rates" will also benefit travellers, as the price of calls from abroad continues to fall.

An e-mail from a laptop in Mikonos to Harlow in the UK costs practically nothing, however I remember 7/7 when terrorist bombs went off in London and suddenly there was no phone signal for my mobile and I wasn't able to reach my wife or son and the panic that I felt until I finally got through about an hour later, on a land line. Imagine if a serious eruption takes place and all communication including land lines, mobiles, satnav, broadband, dial up, in fact if you are stuck in The Seychelles when all this kicks off, I would rather be at home with wet towels around my doors trying to stop the poisonous gases from entering my house, than not knowing where or what was happening to my loved ones. It may be that there are some land lines still working, however another modern trend in the UK, especially amongst the younger generation is to only have a mobile and not to have any land line at all, so there I am in The Seychelles having waited three hours in a queue to make my five minute phone call and only to realise that my son or daughter at Uni in Manchester only has a mobile phone which doesn't work because of rolling power cuts in the UK.

This problem, I personally think will be selective, it may cause an overload on the mobile networks or power cuts will get the networks closed, however broadband on wifi servers may survive as the phone lines are not served by mains power from the National Grid, they have their own very low voltage supplies and might survive a little longer than land lines and mobiles. I do feel though that the Government from its snug bunkers in the Home Counties will make it a priority to get communications back on track as quickly as possible, as without communications, nothing at all can be organized and our civilized way of life would fairly quickly fall apart without communications including TV signals. I would imagine the Government has back up and stand by generators to keep at least the BBC News Channel going as well as emergency radio broadcasts. The public have to be calmed down and told to stay at home and await help or to congregate at the local village or church hall or community centre and supplies will be forthcoming, however don't hold your breath, you may have to wait a long time to receive any help from the Government at the beginning of the crisis.

4. Surviving Abroad During the Emergency

As mentioned in my short story earlier, some travellers will attempt to get home to try and care for loved ones, friends and family, some will decide to stay put in the resort where they were staying at the time the emergency was announced.

Considering the almost impossible trek that you might have to make to get back home, only to find that you are stranded at Dover or Portsmouth without any way to move any further, my advice is, that unless your journey is absolutely critical, to stay put in your hotel. There is a good chance that the computers controlling credit cards might pack up due to power cuts or lack of staff to tend the computers, it might be advisable, if you can afford it, to try and pay for a month in advance, just in case the emergency goes on for some time and you might later not be able to pay the hotelier, who then might throw you out. In my short story scenario the tourists visited the British Consul and they made emergency payments to the hotel, however you might be on an Island or area with no British Consul or in an area like Sharm El Sheik on the Red Sea far away from the British Consul in Cairo. Also, as I stated, The British Government does not have unlimited monies, despite a couple of years of rolling printing presses helping with what was euphemistically called "quantative easing," to my eyes, it's just printing more and more money and the rest of the world will then think the pound is worth less and less, the only saving grace for the pound was that most other countries were in a similar or worse mess than we are, but I digress.

Another option, if you are staying in a five star very expensive hotel, is to quietly have a look around town for a more modest and inexpensive hotel, flat or villa and then try and pay them a month in advance, before your Visa card packs up and then move the family to the more inexpensive hotel, please don't give notice to your hotel to quit until you have firstly made cast iron arrangements elsewhere. However notwithstanding what I have just said, if your tour rep states that Thompsons, Butlins, Centre Parks or whomever have made a deal with the hotelier and that all accommodation and food will be provided during the emergency then for heaven's sake stay put, that would be the golden option.

Quite probably, (as in the previous emergency with EJ) many airlines and tour operators found that they did have to provide accommodation to all their trapped customers for the duration of the emergency. You might have to go up to the tour rep or hotelier (who are acting as agents to the tour companies sometimes) and make a fuss and demand confirmation that under the various International Acts including the 1929 Warsaw Convention and other EU acts that they do have a duty to house you and feed you until they can fly or transport you to your final destination. Try not to be too reserved and British about this issue, let me assure you that the Germans, French and American guests will be making a lot of noise and demanding their rights from day one of the emergency.

The other alternative would be to ring your travel insurers (if they are still at their desks) and ask them to organise with the hoteliers, that all your costs and expenses be covered during the emergency. Perhaps if they have moved their call centre to Mumbai in India there might be a good chance of you getting through and some action being taken. Obviously if your tour operator has already stated that they will cover all your accommodation and other expenses, then you won't need to bother your insurers apart from notifying them of a potential claim for losses incurred whilst you were abroad, i.e. missed appointments, cancelled weddings, engagements, other consequential losses incurred by you whilst you are away.

Even if you do have Internet Banking you might not be able to access it due to power cuts or the site being offline, payments might be missed on HP for cars, mortgage, or rent payments and many other consequences of you and your family being trapped in some far off place for a couple of months maybe, whilst the emergency is ongoing. I hope for all our sakes that whatever volcano does erupt, that the eruption is over within a month or so, it does not bear thinking about if the eruption, such as Laki in 1783 went on for, say six months, it would be very uncomfortable to be trapped on some tiny Island resort or desert resort for six months, let alone the logistics of obtaining medicines (see my earlier chapter about this point), fresh clothes, money to buy toiletries, like toothpastes, shampoo and all manner of other small things which are necessary for civilized living today.

If you did leave the resort and headed towards the airport or shipping port then I would imagine the local authorities would construct some form of tent city so you would be leaving your five star hotel for a small army tent and a smelly sleeping bag, creepy crawlies, bedbugs, cockroaches and queuing for an hour for toilet facilities and water and food. It is in these sorts of temporary camps that outbreaks of dysentery and diarrhoea frequently occur and I would imagine other more dangerous diseases such as typhoid and cholera wouldn't be far behind, these diseases thrive on crowded unsanitary conditions which are where you will be in the middle of tent city. I personally would prefer to stay at the hotel. The local authorities might have commandeered school halls or community centres however you will then be sleeping in some sort of open dormitories and I would imagine that local pick pocket thieves would soon relieve you of all your jewellery, watches, cash and other valuables, please heed my advice, all stay together and try and stay in a decent hotel for as long as possible, even if it means going overdrawn at the bank temporarily.

Many parts of my thesis are dependent on what erupts, for how long and what sort of eruption takes place as well as prevailing weather conditions, where the Jet Stream is and the time of year, as there are so many variables involved we will have to try and look at the differing Volcanoes involved and try and see from past eruptions how long they might erupt for and what came out of them. Then when the CNN news story breaks and you see it is Hekla, which is very dangerous or EJ which comparatively was fairly harmless. Then dear reader you can make some informed judgements whether you are abroad or at home in the UK, Europe or the USA. It may be if Laki or Hekla or similar huge strato volcano starts erupting, you might decide to fly off to Australia or

Singapore for a month to avoid any problems, if you can afford it.

You might decide to drive down to Madrid in Spain to see what happens in the UK and Northern Europe and if things look grim then take the ferry from Gibraltar to Ceuta (Spanish Morocco) or Morocco and stay there, there are hundreds of different options available to everyone, you could sweat it out at home having made good provision for a six month eruption, the eruption phase might only be between two weeks and one month, however as discussed earlier in my book the disruption to basic services, food, light & heating, the banking sector, communications might linger on for months to come. What I am trying to do here is to educate the reader and give you as much forward knowledge as possible so when the worst happens most people can make an informed decision as to what to do.

I do realise that some families with babies, elderly or infirm parents and loved ones may not be able to go anywhere and that is the main reason I have written this book, is to give those sorts of family's time to prepare and have a clear action plan as to what to do. We will all be in the same predicament when the worst happens, I am also banking on the British spirit of helping each other in time of need and not to ignore the elderly spinster living opposite your house who will also need your support and attention or the old gent living down the street, many of that generation gave everything in defence of this country during the last war and they should not now be cast adrift and left to fend for themselves. At the very least try and pop over and see how they are occasionally and inform the emergency services that there is an elderly lady or man living on their own at so and so house on your street.

Getting supplies, heating equipment and food and carers over to nursing homes and hospitals will be another logistical nightmare for the Government to try and plan for, let's hope that this emergency does not actually happen for a considerable time, however from browsing news and other websites it does seem that currently, (September 2010) that there is some seismic activity under and around some of the bigger volcanoes currently in Iceland. I have constructed a website called www.icelandicvolcanoes.co.uk which should be up and running by Xmas 2010 and I will try and either give the link or the story in brief of any impending seismic or volcanic activity that I hear about to my readers and post regular updates on the website, so please use it as a tool to know what is happening regarding the Icelandic Volcanoes.

Section 4

How to survive the upcoming Icelandic volcanic eruptions

Section 4

How to survive the upcoming Icelandic volcanic eruptions

Know your enemy, a list of the likely culprits that will cause the next eruption

I am now going to briefly as possible list as many of the Icelandic Volcanoes that I can find out about and a little about their history and what sort of eruptions and the length of eruption from each one. After each description I will make my own brief comments and then, in my opinion give the probability as a percentage out of 100 of the volcano causing a major eruption within 5 - 10 years from now which in geological terms is a fraction of a second. I will also put my danger levels down as I see it, i.e. a Laki major eruption would be an 8, and a Heimaey, strombolian fountain type eruption would be a 3, whereas a Yellowstone Super Volcano would be a 10. Please note that no one can tell you exactly when a volcano is going to erupt, some give very little notice before erupting, others start to grow and the underground magma chamber (a subterranean cavern of molten rock that slowly fills up with molten magma and eventually explodes up through the volcano's neck, called the vent and then over the surface) fills up, sometimes over hundreds of years like Yellowstone, sometimes quicker like Mount St Helens, volcanologists install laser optical stations (tilt meters) on some volcanoes linked to GPS

Satellites, these lasers can detect the slightest movement in the ground and show when a volcano's magma chamber is on the move, then, if there are dramatic indications of movement the volcanologists can then issue a warning through the news broadcasts or through their institutes.

Unfortunately for Iceland and its human population it sits on top of the mid Atlantic Ridge which is over two tectonic plates, The Eurasian Plate and the North American Plate which are pulling apart from each other at the fantastic speed of a couple of inches per year, however this spreading of the plates allows magma to rise under Iceland, which is also known as a hot spot, in this case not a night club in town but an extra hot molten plume of magma rising from deep within the earth's mantle, and regular eruptions have occurred through recorded and prehistoric times, which have actually built up the Island.

If you go to Iceland on holiday and you go on some tours, they will take you to bubbling hot springs and lakes, some of which you can swim in, bubbling mud pools and you can see some of the major active volcanoes across the Island. Reykjavik itself, (the capital city of Iceland) gets its heating directly from the hot magma under Iceland which heats the underground water giving free heating to most Icelanders, this method of heating up water is called geothermal.

What we really need is a pipeline from Iceland to the UK so they can send us free hot water and steam for the UK in return for all the money they owe us for Icesave and the other failed Icelandic Banks, however that is also another story for another time.

Currently there is an alert out about Bardarbunga / Grimsvotn from April 20th 2010. This is one of the large Strato Volcanoes which has a 700 metre deep Caldera and lies beneath the Vatnajokull Icecap.

Recently in 1999 there were 7 separate clusters of quakes in the vicinity of Kolbeinsay Ridge which is part of the localised area of the Icecap.

Previously a huge fissure eruption at Thjolsaihraun produced approximately 21 km3 (21 cubic Kilometres of lava) the largest know Holocene (The period between the end of the last Ice Age and the present day) lava flow on the planet.

I have heard no new news recently, please regularly keep an eye on my website www.icelandicvolcanoes.co.uk and check other related websites to keep in touch with the current situation. Please also note that the opinions expressed below as to the likelihood of a certain volcano erupting and its danger levels are my personal opinion based on my reading of various documents and studying various websites regarding the history of these Icelandic Volcanoes. I am not guaranteeing or warranting any facts or figures as 100 % correct and would advise interested readers to make their own enquiries as to the seriousness or otherwise of any of the scenarios I have described below.

1. Askja

"A view of Lake Oskjuvatn which fills much of the caldera of the Askja volcano, it covers 12km2, this lake was the result of the major 1875 eruption. The whole area is extremely remote, only being accessible by land for a few months of the year."

Style: stratovolcano
Location: Dyngiufjoll Mountain Range (Centre West of Iceland)
Height: 1,510m (4,954 ft)
Last Erupted: 1961

Askja is situated in one of the remotest parts of Iceland, in what is called "The Central Highlands." The name Askja refers to a complex of calderas which are very close to each other. This remote area is within the Dyngjufjoll Mountains which are 1,510 m (4,954 ft) high.

The first anyone knew about Askja was when a huge eruption occurred in 1875. The Eastern part of the country had such heavy ash fall that many cattle died and the crops were ruined.

The eruption was so powerful that volcanic ash (tephra) was blown on the winds and reached Norway, Sweden and parts of Denmark. Due to the continuing eruptions and the poisoning of the land this eruption caused many of the local inhabitants of Eastern Iceland to migrate from Iceland.

Askja previously erupted about 9000 BCE just after the last Ice Age and ash from this eruption has been found from Ireland, Norway and Sweden. Askja last erupted in 1961 so should still be considered active. There are a whole series of calderas in the area, the outer caldera is about 50 km2, and its height is approximately 1,100 m. There is a lake called Oskjuvatn which fills a smaller caldera formed from the 1875 eruption, it is about 12 km2 and used to be warm, however currently it is frozen for much of the year.

Viti is another small caldera N E of Oskjuvatn, it contains a geothermal lake of warm sulphurous opaque blue water which tourists can swim in, however they are warned to be very careful as a build up of carbon dioxide can cause swimmers to pass out and drown, perhaps not the place to take granny for a summer swim.

In June 2010 a volcanic expert, Hazel Rymer indicated that underground seismic activity was on the increase at Askja and that an eruption could be imminent. Scientists do not think that (EJ) Eyjafjallajokull is responsible for current activity at Askja, scientists continue to monitor Askja

My Current Opinion

This is a big volcano and one that we do have to keep an eye on, It has caused exactly the sort of scenario that we have mentioned earlier in our book. We would need to see further Laser tilt readings to see if the volcano is rising due to its magma chambers filling again before making an informed guesstimate as to when it will erupt again.

Danger Level: 7 (out of 10)

Eruption Level: 65 % probability of a major eruption in the next 10 years.

2. Bardarbunga (Warning Level 1 Risk)

By kind permission of Mr Oddur Sigurdsson jardfraedingur (geologist) Icelandic Meteorological Office, Reykjavik, Iceland

"An aerial view of Bardarbunga, showing a huge glacier sitting on top of a strong massive and powerful stratovolcano, part of Iceland's largest volcanic system nearly 200 kms long and up to 25 kms wide. The volcano is covered in ice hiding her massive glacier filled caldera."

Style: Large stratovolcano
Location: Under the Vatnajokull Glacier
Height: 2,009 m (6,591 feet)
Last Erupted: 1910

Bardarbunga is an extremely dangerous stratovolcano that we have mentioned already at the beginning of this chapter. It is Iceland's biggest volcanic system nearly 200 kms long and 25 kms wide. You cannot see this volcano as it is hidden under a huge glacier currently.

Bardarbunga is situated in the remote Central Highlands of Iceland far away from most inhabited areas. Icelanders however do have a fearful respect for this monster of a volcano.

The main Caldera is about 70 km2 and up to 10 kms wide and 700 m deep. The lip of the crater is about 1850 m high and is completely submerged in ice as it is partly under a glacier. There are many strata of ash built up around the crater which shows that Bardarbunga has been very active over a long period of time. The Gjalp eruption of 1996 showed the scientists that there may be a linkage between Bardarbunga and Grimsvotn.

A large earthquake measuring 5 on the Richter scale is thought to have ignited an eruption at Gjalp.

There has been regular seismic activity under and around Bardarbunga for some years now which could be an indication of underground magma chambers filling up prior to another big eruption soon, scientists cannot yet agree as to what might happen here and any timescale for the future eruption of Bardarbunga, although it is widely thought that many of the volcanic systems under Iceland may be linked at deep levels under the earth's crust. Previously there appears to have been large eruptions where the magma moves quickly under the volcano to the southwest of the caldera complex.

Bardarbunga is definitely one of the "big boys" when it comes to mega large eruptions. BB (Bardarbunga) appears to erupt fairly regularly every 250 – 600 years. It has been deduced by scientists that one of the largest lava flows in Iceland and in fact on the whole planet occurred about 6,500 BCE, the flows covered between 21 – 30 cubic kms and covers an area of land of approximately 950 sq kms.

It appears that many eruptions have occurred during the Holocene period. After humans inhabited Iceland in the ninth century the records show that Vatnaoldur (part of the BB system) erupted in 870 ACE and Veidivotn erupted in 1480 ACE, both of these mega eruptions would have had a wide scale effect on habitation by humans and livestock and wild animals on Iceland and should these size eruptions occur today, they would have a major direct effect on everyone's lifestyle in Northern Europe, Scandinavia and surrounding areas.

It appears that in recent times that regular small eruptions have occurred beneath the actual glacier, probably in the north east of the crater or actually inside BB. These eruptions seem to follow a pattern, there were several eruptions in the glacier between 1701 – 1740, other small eruptions in the nineteenth century and continuing regular seismic activity including regular earthquakes have been construed by the scientific community that BB will erupt again in a big way in the near future, whether this means in one year's time or twenty years time is very difficult to calculate currently.

My current opinion

Bardarbunga is definitely one of the top five stratovolcano's that we have to watch very closely. The sort of mega eruptions that BB could make include large volcanic fissure eruptions that seem to occur every 5000 – 8000 years south east of BB. These eruptions could be huge and actually change the face of Iceland and cause mayhem with Iceland's hydro electric power supply grid. Ash Clouds from any major BB eruption would definitely cause freezing winters and poor summers in Europe and cancel all flights to Europe for many months amongst other life changing matters discussed earlier in my book. I will keep a close eye on BB and post updates on my website. www.icelandicvolcanoes.co.uk

Danger Level: 8 (out of 10)

Eruptability Level: 55 % probability of an eruption in the next 10 years.

3. Eldfell (Heimaey – Surtsey)

"A view of Eldfell, which is situated on the small island of Heimaey, to the South of Iceland, it erupted without warning on 23rd January 1973 to eventually become 200 metres high, nearly destroying the fishing port on Heimaey. The volcanic cone suffers from wind and sea erosion, the locals are now planting trees to stop the erosion continuing."

Style: Composite Volcano (stratovolcano)
Location: On the Icelandic island of Heimaey, part of the Vestmannaeyjar (Western Islands) situated off the Southern tip of Iceland
Height: 200m (656 ft)
Last Erupted: 1973

This Composite Volcano is located on the island Heimaey situated off the Southern coastline of Iceland. The volcano erupted without warning on January 23rd 1973. Eldfell means "Mountain of Fire" in Icelandic.

The eruption caused major problems for the Island as the huge ash fall and then the flows of lava nearly destroyed the fishing town of Heimaey. There was discussion of permanently evacuating the inhabitants to the mainland. The authorities then devised a huge pumping operation and managed to pump huge amounts of sea water onto the advancing lava flows to save the town at the last moment.

The history of the area was that the Islands were settled in 874 ACE, originally by escaped Irish slaves, although the islands have only a poor water supply the Island flourished as being one of the only fishing ports along Southern Iceland and the fisheries off the island are teeming with all types of fish which have sustained Heimaey for many centuries, there was no evidence of any volcanic activity until the volcanic island of Surtsey appeared above the waves in 1963, although there are thought to have been other off shore underwater eruptions in 1637 and 1896.

The eruption of 1973, although it was preceded by some earth tremors was very sudden, at 01:55am on 23rd January 1973 a fissure opened up 650 ft east of Kirkjubaer (Church Farm). The fissure grew from 300 metres to 2 kms, crossing the island from one side to the other and submarine volcanic activity occurred off both sides of the fissure under the sea for some time afterwards. Fantastic lava fountains 50 – 150 m high were encountered along the whole length of the fissure, however most of the volcanic activity soon became centred on one large vent about 0.8 km north of the old volcanic cone of Helgafell just outside of the eastern edge of the town of Heimaey.

The eruptions continues and Eldfell grew to 100 m (330 ft high) and the strombolian eruptions caused a massive delta of lava to flow north and east and started to encroach upon Heimaey's only harbour. The ash cloud column rose 9000 m (30,000 ft) nearly as high as the Tropopause.

The authorities then decided to evacuate all the 5,300 people from Heimaey as the eruption continued, they were temporarily taken to the mainland. The eruption continued and houses collapsed under the weight of tephra, all the livestock had to be slaughtered as they could not evacuate all the cattle and sheep and other farm animals.

A few brave souls stayed in the town to try and keep the roofs of houses clear of tephra, however as much as 5 m of tephra fell across the island. Some houses were destroyed by lava bombs and by streams of red hot molten lava advancing through the town, it seems like a picture from Dante's infernos of hell.

By early February the tephra had abated somewhat, however the lava flows continued and submarine activity severed the electric supply from the mainland and the water supply from the mainland. The lava was now flowing into the harbour and as Heimaey was responsible for about 25 % of Iceland's fishing revenue the authorities began to get really worried. The lava flows covered much of the Island and created about 2 sq kms of new land on the island, the lava flow was 100 m (233 ft) thick and spread across much of the Island and was starting to encroach further across the town itself.

The authorities devised a scheme to lay water pipes into the town and pump huge amounts of sea water onto the advancing lava flows to stop the harbour and town being destroyed and although it was a dangerous and difficult job this action started to succeed and the lava flows started to slow down and stop.

On one occasion a large piece of the crater wall broke away and was being carried along in the lava towards the harbour the locals called these chunks of lava rock "the wanderers" and only by expert spraying of sea water under the advice of Prof Porbjorn Sigurgeirsson on board the dredging boat Sandey were these huge chunks of crater prevented from destroying the harbour.

By March of 1973 the eruption had died down from an initial flow rate of 100 cu m per second (3,500 cu ft per second) to 5 cu m per second (180 cu ft per second). Shortly before the eruption ceased a tilt meter, situated at a height of 1150 m (3750 ft) which had been measuring ground deformation throughout the eruption measured a tilt towards the crater which implied that the underground magma chamber was nearly empty and was perhaps collapsing in on itself at the end of the eruption. In total about 0.25 cu km (0.06 cu mile) of material had erupted from the crater. About 2.5 sq km (1 sq mile) of new land was formed and added to the island increasing its area by 20 %. The lava flows had narrowed the harbour's entrance, however fishing vessels could still easily navigate through the harbour entrance and the new lava walls were similar to a breakwater which helped to shelter the harbour from Atlantic storms which are common in these latitudes.

Most of the Islanders eventually returned to Heimaey and the local authorities managed to use the cooling lava flows as geothermal heaters to produce hot water and heat and some power from the cooling lava, nowadays Heimaey's fishing fleet provides about 30 % of all fish in Iceland and the islanders are planting grass seeds around the base of the volcano to stabilise its flanks against erosion by the elements.

My current opinion

Although Heimaey's story is a fascinating story, as is Surtsey's, which was the story of a submarine volcano bursting through the cold waters of the Atlantic and forming a completely new island in 1963 (I remember the story well although I was only a wee bairn in those days). I do not think that these strombolian style eruptions are as dangerous to us in the UK and the rest of Europe as the likes of BB, Katla and the formidable Laki. These sorts of eruptions, as they are fairly free flowing and include the fountain style strombolian eruptions of lava projection and are not as explosive as the stratovolcano's buried under a km of ice, as many of the other more dangerous volcanoes are in Iceland and for that reason I have downgraded their danger, so even if they did erupt it would not be as dangerous as many of the other volcanoes mentioned in the pages of this book.

Danger Level: 3 (out of 10)

Eruptability Level: 40 % probability of a further eruption in the next 10 years

4. Eldgja

"A photo of Eldgja , the largest volcanic canyon in the world, 270 metres deep and 600 metres wide, it created one of the largest flows of lava in historic times in the year 934 ACE, an estimated 18 km3 of lava flowed out of the earth's mantle."

Style: Volcanic Canyon (flood Basaltic style)
Location: In the south of Iceland situated between Landmannalaugar and Kirkjubaejarklaustur.
Height: 800 m (2,625 ft)
Last Erupted: 934 AD

Eldgja is a volcanic canyon or valley and is linked to the volcanic system that includes the dangerous stratovolcano called Katla. Eldgja means "fire canyon" in Icelandic.

Eldgja is the largest volcanic valley in the world at 270 m deep and 600 m wide at its greatest width. Porvaldur Thoroddsen was the first person to fully understand the true nature of this volcanic canyon In 1893.

It appears that about the year 934 the largest flood basalt (freely flowing basaltic lava flows) eruption occurred in the historic era of time, an estimated 18 km2 of lava poured out of the earth's interior. In earlier era's of time, millions. of years ago, basaltic lava flows of massive scale, called the Siberian Traps in Russia and the Deccan Traps in India, caused massive changes in the consistency and makeup of the earth's atmosphere by pumping huge amounts of carbon dioxide and other gases into the atmosphere, that all life on earth nearly died out, this era marks one of the biggest mass extinctions know on the planet. We ignore mother nature's internal movements under our feet at our own peril and although had the "Siberian Traps" flood basalts happened today we probably couldn't do much about it, it is always wise to know what might be out there waiting to bite us, as they say forewarned is forearmed.

My Current opinion

Eldgja, from what I can gather seems fairly quiet over the last 1000 years or so, however as briefly mentioned above when flood basalt lave flows occur they can release some greenhouse gases such as CO_2 and other gases which could speed up climate change on earth today. I have heard reports that climate change is responsible for the thinning out of many glaciers in Greenland and also possibly in Iceland too, this reduction of weight over many of these strato volcanoes and other volcanic systems could act rather like a champagne cork being taken out of a bottle, once the cork is released, then a massive amount of renewed volcanism in the Icelandic area might soon follow behind, more food for thought.

I personally do not think that Eldgja could erupt very soon, there are no reports of seismic activity that I am currently aware of, however as with previous volcanoes I will keep an eye on it and post any changes I see on the website.

Danger Level: 2 (out of 10)

Eruptability Level: 10 % probability of eruption in the next 10 years.

5. Esjufjoll

By kind permission of Mr Oddur Sigurdsson jardfraedingur (geologist) Icelandic Meteorological Office, Reykjavik, Iceland
"An aerial view of Esjufjoll, a massive subglacial volcano hidden for the most part under the huge Vatnajokull icecap. About 40 km2 of the volcanic caldera is covered by the icecap, however part of the south eastern flanks of the volcano are now exposed, a sub glacial eruption was reported in 1927 and caused some local flooding."

Style: Stratovolcano
Location: In the SE part of the Vatnajokull icecap, north of Oraefajokull volcano
Height: 1,760 m (5,774 ft)
Last Erupted: 1927

This volcano is buried beneath an icecap, the volcano consists of the Snaehetta central volcano and a large caldera. Parts of the SE side of the volcano are exposed in NW – SE ridges. The caldera is 40 km2, making it a very large volcano. There has been some minor activity in 1927 and the smell of sulphur has been detected as well as occasional ash fall, although since the end of the last ice age, about 10,000 years ago, not much activity has been noticed, recently earthquake swarms were detected in 2002 that could show magma flows are on the move as the ice cap slowly melts due to global climate change.

My current opinion

I personally do not think that we currently have much to fear from this slumbering giant, I think there are many much more dangerous Icelandic volcanoes out there to worry about. My only reservation in saying the previous statement is that, sod's law, the one volcano that I say is harmless will be the one that suddenly blows up out of nowhere.(lets all hope that my initial prediction is the correct one).

Danger Level: 2 (out of 10)

Eruptability Level: 15 % probability of an eruption in the next 10 years.

6. Eyjafjallajokull (Warning Level 1 Risk)

By kind permission of Mr Oddur Sigurdsson jardfraedingur (geologist) Icelandic Meteorological Office, Reykjavik, Iceland

"An aerial view of the Eyjafjallajokull icecap, slumbering fitfully until the next huge eruption bursts through the protective icecap."

"Eyjafjallajokull erupting through its icecap, is the name of the icecap and the actual volcano that erupted in March to April 2010. There are indications that subterranean links from the magma chambers might be linked to the much more powerful Katla volcano which is only about 30 kms away."

Style: stratovolcano
Location: Situated on the Southern coast of Iceland north of Skogar and to the west of Myrdalsjokull.
Height: 1,666 m (5,466 ft)
Last Erupted: February – May 2010 (and continuing)

This is the volcano that all the newsreaders and reporters round the world hate, as no one can pronounce it correctly except perhaps Icelandic reporters, although it's very easy to learn, Eyjafjallajokull is pronounced "eija, fjatla, joekl", see simple's (as the meerskat advert goes). This in effect is the one volcanic eruption that focussed the world's attention and my attention in particular onto the wonderful, if dangerous kaleidoscopic world of Icelandic Volcanoes.

In English EJ, (as we shall call it for now) is Icelandic for "Island Mountain," the ice cap, more or less completely covers the central caldera. EJ erupts with horological (clockwork) frequency since the beginning of the Holocene period, in the early 19th century between 1821 to 1823 it erupted and again this year.

The crater is 3 – 4 kms (1.9 – 2.5 miles) in diameter, open to the north, the crater rim has three peaks, Guonasteinn, Hamundur and Gooasteinn. The south face was once situated right on Iceland's Atlantic coastline from which over many millennia the sea has retreated some 5 kms (3.1 miles) as various eruptions have produced new lava which has increased the valleys between EJ and the sea, this former coastline now consists of massive cliffs with spectacular waterfalls cascading over the edge, of which the best known is

Skogafoss. The coastal plain between the mountain and the sea is called Eyjafoll.

The composition of EJ is mainly basalt to andesite lavas, most of the historical eruptions have been explosive, however fissure vents occur on several sides of the volcano.

EJ is linked to a large magma chamber underneath the mountain which is linked to the main tectonic spreading ridge called "The Mid Atlantic Ridge," EJ is also linked possibly through various subterranean magma chambers and passageways, that we currently know very little about, to its big neighbour, Katla (which we will discuss in depth later in our book), however the salient point to remember here is that historically and geologically when EJ erupts, normally fairly soon afterwards Katla blows its top and that is the $ 65,000 question, when and if will Katla erupt and due to the size of Katla, the EJ eruptions will seem like a Sunday walk in the park compared to what Katla could do, more on this later in the book. EJ is also linked to Eldfell on Heimaey (which we have already discussed).

Eyjafjallajokull has previously erupted during recorded history in 920 ACE, 1612 and from 1821 – 1823, when it caused a glacial lake flood called a jokulhlaup to occur. Its recent eruptions were in March 2010 when a brief evacuation of 500 people occurred. However the April 14th eruption was 20 times more powerful than the March eruption and as we know, has caused severe disruption to air traffic from USA and Canada to Europe and throughout much of Europe. For about two weeks there was no flights, which cost the airlines billions of pounds in lost income and trapped travellers all over the globe, some travellers took five weeks to get back home and many spent thousands of pounds to try and get home. Don't forget, this EJ eruption is probably one of the most tamest and peaceful compared to many of the other volcanoes that we have been discussing in this book.

The 1821 to 1823 eruptions

In the first eruption of 1821 it has been noted that contained within the eruptive ash is a large percentage of fluoride which in large amounts can cause bone damage in both farm animals and humans. This earlier eruption also caused some localised flooding in the nearby rivers of Markarfjot and Holtsa.

From June to August of 1822 a further series of explosive eruptions occurred. The ash cloud rose many thousands of feet into the air above Iceland and ash rained down on the northern parts of Iceland, in Eyjafjordur and also in the southwest on the peninsula of Seltjarnarnes near to Reykjavik. Between August and December 1822 the eruptions seemed to die down somewhat, however many cattle and sheep in the Eyjafjordur region died from some sort of gas poisoning during this period, modern day scientists have analyzed what this poison gas might have been and have deduced that it was fluoride gas poisoning. This is just one of many poisonous gases which can emanate from volcanoes, especially ones under icecaps which can react with the ice and snow in the icecap to make a deadly cocktail mixture of gases, many of which are poisonous to humans and animals alike. Some small glaciers partially melted due to the eruptions and the run off flowed into the river Holtsa, a larger glacier melt flooded the plains near the river Markarfjot, possibly also in 1822 or thereabouts.

The 2010 eruptions

In December 2009 thousands of small earthquakes were detected in the vicinity of EJ, mostly in the range of 1 – 2 on the Richter Scale, these earthquakes occurred 7 – 10 kms (4.3 – 6.2 miles) underneath the volcano. On 26th February unusual seismic activity together with a rapid bulging and deforming of the earth's crust was registered by the Meteorological Institute of Iceland. This was the first hard piece of evidence that magma was pouring from the Earth's Mantle through the spreading fault and into the magma chambers directly underneath Eyjafjallajokull (EJ to its friends), the upwards thrusting pressure from this quickly filling magmatic chamber, a typically Icelandic Volcanic style of vulcanology, was causing a huge crustal displacement at The Porvaldseyri Farm. The seismic activity increased between 3 rd to the 5th March with nearly 3000 small earthquakes coming from the centre underneath the volcano registered on seismic reading equipment.

The first eruption into the atmosphere occurred on March 20th 2010 about 8 kms (5 miles) east of the main crater, on Fimmvordukals, the high ridge between EJ and the Myrdalsjokull Icecap neighbouring EJ. This was the first eruption and was a fissure eruption which erupted directly into the atmosphere and was not hidden under the icecap. It was smaller in scale than most of the scientists had predicted. The fissure did little damage originally, as this is a very sparsely inhabited area apart from a few farms and small hamlets. Only a hiking route was split by the fissure eruption between Skogar, south of the ridge, and Porsmork to the north.

On April 14th 2010 EJ again erupted after briefly lying dormant again, though this time it was from the top crater directly under the main glacier, this caused fairly immediate glacial flooding, known as "jokulhlaup," which poured down local rivers causing the immediate evacuation of 800 local dwellers of the area. The main reason for the drama that then unfolded is that the chemical makeup of the magma reacted with the melting glacier causing the sort of ash clouds that are a danger to aircraft in that they go up many thousands of metre's into the stratosphere, possibly up to the tropopause and can be ingested by jet engines causing engine failure and resultant damage to the aircraft. There is also the possibility of further fluoride poisoning of humans and livestock taking place and even the remote possibility of a mixing of chemicals and gases to produce components of the "Laki Haze" of the eighteenth century which killed thousands of people across Europe and the USA.

The volcanic eruptions at EJ were explosive in nature due to the water getting into the fissure vents of the volcano. This eruption was about 20 times more powerful than the previous eruptions in Fimmvorduhals. This eruption threw volcanic ash and debris up to 3 kms into the atmosphere which led to the skies over most of Europe being closed to air traffic for 6 days from April 15th and again in May of 2010. Many airlines lost millions of pounds in lost revenue and compensation claims that are still rumbling on through various insurance companies negotiations. It is estimated that the airlines losses could mean that some of the airlines involved actually will make a pre tax loss on business activities in 2010, and this was only a very brief closure of the skies for about 6 days, could one imagine if Katla or Laki or one of the other "big boys" erupted, what would happen? It would possibly mean that several airlines and maybe national carriers would end up in receivership, or going bust, as they just wouldn't have the spare capacity to keep all their planes on the ground for week after week of large amounts of high level and low level ash carpeting much of Europe and parts of North America. Many other businesses would be severely damaged, as mentioned before in this book many thousands, maybe hundreds of thousands of travellers would be stranded all over the world unable to return to their home country for maybe months on end.

The eruptions also created violent electrical storms. On 23rd May 2010 the London Volcanic Ash Advisory Commission advised that for all intents and purposes that the current spate of eruptions had ceased, however they were continuing to constantly monitor the volcano. I will of course be regularly checking their website and the various Icelandic websites and keeping my special Icelandic Volcano website fully updated at all times. Currently the volcano continues to have a few small earthquakes on a daily basis with many local and international volcanologists closely monitoring the volcano's every movement. The problem with many of the Icelandic Volcanoes is their lack of warning before erupting, this maybe because Iceland is located over a hotspot between the earth's crustal plates and upwelling magma seems to travel very quickly through the underground fissure's and pipes and seems to fill the underground magma chambers very quickly. This is because the chemical composition of some lava's like flood basalt's are quick flowing, however others again depending on the chemical makeup are slow moving, no two volcano's lava flows are exactly the same therefore. However as a generality the Icelandic magma and lava does seem to move faster than many other lava's and magma's and many of the historic eruptions seem to have occurred without much warning beforehand, it must also be noted that a small eruption under an icecap or glacier may not go noticed at all.

However when the eruption is big enough and it burst through the icecap or glacier then the volcano is already up and running and the first anyone seems to know is that a huge ash cloud is rising through the atmosphere and giant glacial melt floods are causing havoc on the ground, by then all anyone can do is evacuate the local population and close any airspace that might be affected by the windblown ash clouds. If the wind is blowing towards the east or south east, will mean that the UK and EU will be affected with maybe only 48 – 72 hours notice of an impending catastrophe, not much time to prepare for the worst. However, for those who prudently purchased this self help awareness book and took some precautions against 3 – 6 months (or even a years) problems, then the worst effects, e.g. lack of food, heating, water and medicines should not cause undue problems, however it is my assumption that sadly many citizens will not have had the foresight to have thought ahead and they will be thrown upon the forward planning of the Government to clothe, feed and keep them warm for up to a year. I, personally, would rather rely on my own wits than to be reliant on the authorities currently. I fervently hope that the advertising and publicity surrounding the launch of this book will jolt those Westminster Mandarins responsible for the public's safety into a massive spending spree of food, water, medicines, emergency housing, tents, vehicles, fuel and clothing and Campingaz canisters to be stockpiled in secure locations, so that if the worst happens, then there will be ample distribution and those like myself who have planned ahead won't be totally reliant on our own devices.

Unfortunately I do sympathise with the Government at a time when up to 40 % reduction in public spending across the board is being announced, to have to spend the UK's precious resources on the above might be a difficult decision to take, let me assure those ministers involved that to not spend these monies and then the worst happens would be a much more terrible outcome and disaster for any administration than to have to rotate mountains of cans of food every six months and similar matters. I suppose the only upside if the Government goes on a spending spree is that Tesco, Asda, Morrison, Lidl and Sainsbury's shares will zoom up making the shareholders some extra pocket money.

The volcanologists only consider a volcano to be dormant when no activity has occurred for three months or more.

(Evil twins) Eyjafjallajokull and Katla

Although we will be discussing Katla in more detail later, I do have a duty to mention the sinister relationship between these two Icelandic subglacial stratovolcanoes. EJ (the smaller and less dangerous of these two volcano's) lies 25 kms west of another much more powerful and dangerous volcano called Katla, which lies under the Myrdalsjokull ice cap.

Katla is many times more active then EJ and is well known for its powerful subglacial eruptions and its huge magma chamber underneath the volcano. Every time EJ has blown its top, i.e. 920, 1612, 1821 – 1823, Katla has a small time later also blown its top off. Currently volcanologists have not seen any recent unusual activity such as ballooning of the crust due to magmatic displacement or any earthquake swarms during the current EJ activity, although geologists have seen the general de stabilisation of the Katla region since 1999. Many geophysicists think that the EJ eruption may cause the imminent eruption of Katla which would cause a massive amount of flooding on Iceland and send up a monster sized ash cloud which would be one of the nightmare scenarios we have been discussing in this book. On 20th April 2010 the President of Iceland stated, "the time for Katla to erupt is coming close..... we, (Iceland) have prepared...it is high time for European governments and airline authorities all over the world to start planning for the eventual Katla eruption."

As of August 2010, volcanologists are still watching and monitoring both Katla and EJ as eruptions of Katla normally follow EJ by a few months, especial checks are being made to see if any small earthquake swarms are emanating from the Katla EJ area, my website www.icelandicvolcanos.co.uk will be updated as soon as I know anything.

My current opinion

I have been perusing several volcano websites and note that many volcanologists are getting very nervous and edgy about the EJ and Katla link, in one website I browsed there was a schematic of all the underground magma chambers from both volcanoes and at various depths some of the chambers are quite close together, many scientists theorise that the upwelling of magma into EJ starts magma moving into Katla's magmatic chambers as well. The main difference though in the eruptive cycle of these two volcanoes is that Katla has a much more explosive mix of chemicals, including rhyolite's and basalts, amongst other mineral compounds and is much like Mount Pinatubo in the Philippines and when that volcano erupted in 2006 it lowered temperatures around the world for about two years.

I personally think that definitely within 5 years we will see some more activity from EJ and quite probably a major blow up from Katla, the Icelandic President definitely thinks so and I am sure many of his countrymen also think so, what do they know that we in Europe don't know, that's what worries me.

Danger Level: (EJ & Katla) 8 (out of 10)

Eruptability Level: 80 % probability of an eruption in the next 5 – 10 years,

7. Grimsnes

"A photo of Grimsnes showing the caldera filled with a small lake, this is one of the smaller fissure style volcanic systems in Iceland."

Style: Fissure or crater row volcano
Location: Located in SW Iceland, SE of Thingvallavatn Lake stretching across the Reykjanes Peninsula.
Height: 214 metres
Last Erupted: 3,500 BC

Grimsnes is a relatively small volcanic area situated across the Reykjanes Peninsula. The lava fields cover a total area of about 54 km2 (21 sq miles). The largest lava field is the Seydisholar – Kerholahraun field which covers 23.5 km2 (9.1 sq miles). Other areas of lava field are the Tjarnarholahraun field which covers 11.9 km2 (4.6 sq miles) and various other smaller lava fields.

My current opinion

This appears to be an ancient dormant lava field I doubt if this area will cause any trouble in the foreseeable future.

Danger Level: 1 (out of 10)

Eruptability Level: 5 % probability of an eruption in the next 10 years.

8. Grimsvotn (Warning Level 1 Risk)

"An aerial view of part of the massive Grimsvotn fissure volcanic system erupting in 2004, this is part of the same system that holds the extremely dangerous volcano called Laki. The volcanic system is mostly covered by the Vatnajokull glacier and its ice cap. Most of the Grimsvotn eruptions are subglacial and produce occasional flooding of the nearby river systems when the volcano melts some of the ice of the glaciers."

Style: Volcanic Fissure System (linked to the Laki volcanic system)
Location: In the Highlands of Iceland at the North-western side of the Vatnajokull Glacier.
Height: 1,725 m (5,659 ft)
Last Erupted: Nov 2004

The Grimsvotn Lakes are situated at the North Western side of the Vatnajokull Glacier and are currently covered by an ice cap. Grimsvotn is a part of the huge earth climate altering volcano called Laki (which we will discuss in greater depths later in our book). Grimsvotn did erupt at the same time as the infamous poison gas cloud eruption of Laki in 1783 – 1784, however Grimsvotn continued to erupt until 1785. Most of the volcano lies below a glacier, therefore most of the historic eruptions have not been recognised, as the eruptions were subglacial.

As we mentioned before subglacial eruptions can cause huge immediate flooding outbursts called "Jokulhlaup's," in Grimsvotn's case the volcanic caldera or crater could quickly fill with water which could lift the overlapping ice cap allowing massive flooding of the surrounding areas to take place, therefore scientists continue to closely monitor this volcanic system very closely. A large eruption did occur in 1996 and a huge Jokulhlaup (volcanic induced flood) did occur about two weeks after the eruption and washed part of the main Icelandic ring road away, luckily no one was injured in that incident.

Recent eruptions occurred in 1998, luckily no flooding occurred and again Grimsvotn erupted in November 2004, however this time there was some disruption to

European Airline Traffic from the ash cloud which travelled as far as Mainland Europe, no glacial outburst flooding occurred during this eruption either.

An interesting phenomenon occurred, in that bacteria were found in the volcanic subglacial lake where there is very low densities of oxygen, it is leading some scientists to surmise that there could similarly be some bacteria that could survive on the planet Mars under similar conditions and it is now planned that when the next Mars Space Mission occurs, that the exploring teams will use the same search criteria as on Earth.

My current opinion

Grimsvotn is linked to the deadly Laki volcanic system, which in itself should start alarm bells ringing everywhere. I personally feel that this system needs to be closely monitored together with the Laki system as they will probably go off together in a big way at some time in the not too distant future. This volcano seems to be still very active and if it was pouring out ash clouds and closing down airspace as recently as 2004, it has to be placed firmly on the danger list of volcanic systems to watch over the coming months and years. There is no doubt in my mind that someday soon our TV screens will be full of the dreaded words, "Laki and Grimsvotn have erupted and a huge ash cloud of terrifying size is heading towards the UK and Europe." I hope I am wrong, however history has taught us one basic fact, that if an event has happened before it can definitely happen again. Keep an eye on this one on my website www.icelandicvolcanoes.co.uk.

Danger Level: 8 (out of 10)

Eruptability Level: 80 % probability of an eruption in the next 5 years.

9. Hekla (Warning Level 1 Risk)

By kind permission of Mr Oddur Sigurdsson jardfraedingur (geologist) Icelandic Meteorological Office, Reykjavik, Iceland

"An aerial shot of the infamous Hekla, one of the most dangerous of all the Icelandic Volcanoes. Hekla once called, "the gateway to hell" in the middle ages. It has erupted with cataclysmic eruptions on numerous occasions in historic times, today fitfully slumbering at present, one hopes it will carry on sleeping for some time to come."

Style: Active Fissure Strato Volcano
Location: South Eastern Iceland
Height: 1,488 m (4,882 ft)
Last Erupted: February 28th 2000

Hekla, like Laki, Grimsvotn Katla and Bardarbunga is one of the most dangerous and active volcanoes on Iceland and in my opinion, if Hekla blows, we in Europe will very soon know all about it. Historically more than twenty major eruptions have been recorded since the year 874 ACE when Icelandic recorded history began. Hekla was known in medieval times as "the gateway to hell," not exactly one of the most charming epitaphs to give a naturally occurring feature.

Hekla is situated on a volcanic ridge about 40 kms (25 miles) long. However the really active part of this volcano is a fissure about 5.5 km (3.4 mi) long named Heklugja, and this feature is actually considered to be the actual volcano. The whole ridge looks something like an overturned ship with the jutting out keel being in fact a series of craters, two of which are the most active.

Iceland is covered in tephra from all the various volcanic eruptions over the millennia, however it is possible using analytical techniques based on the chemical composition of the ash that volcanologists can determine from where and when an eruption occurred, i.e. the smoking gun after an eruption. About 10 % of all tephra (volcanic ash) over the last thousand years has come from Hekla which equates to five cubic kilometres of ash, which in my opinion is a huge amount of ash. Hekla has also produced eight cubic kilometres of lava which is one of the largest amounts of lava produced from any volcano on Earth in historic times.

Hekla in Icelandic is the word for a short hooded cloak, which may be a reference to the clouds which frequently seem "to cloak" Mount Hekla. A Latin source named the mountain as "Mons Casule." After the major eruption of 1104 ACE, stories deliberately spread through Europe by Cistercian Monks implied that Hekla was actually in reality the real "Gateway to Hell" on Earth. The Cistercian Monk, Herbert of Clairvaux wrote in his De Miraculis that;

"The renowned fiery cauldron of Sicily,(possibly Mount Etna) which men call Hell's chimney.....that cauldron is affirmed to be like a small furnace compared to this enormous inferno."

Herbert of Clairvaux, Liber De Miraculis 1180.

A poem by the monk Benedeit from circa 1120 ACE about the voyages of the Irish monk, Saint Brendan mentions Hekla as the prison of Judas.

St Brendan, was an 8th Century monk who set sail from Ireland via Iona with a band of hardy monks in an animal skin boat sewed together and with a sail, it is thought St Brendan got as far as Iceland and some think maybe even further West to Greenland. There was one occasion that the monks encountered a small Island and went ashore only to find out that it was a huge whale which was resting before slipping below the sea again and the monks had to make a hurried dash to get back on board their boat before their "island" sank again into the seas.

In the Flatey Book Annals writing of the year 1341 ACE, that the eruption (from Hekla) that people saw large and small birds flying in the mountain's fire, which were taken to be souls. In the 16th Century Caspar Peucer

242

wrote that the Gates of Hell could be found in "The bottomless abyss of Hekla Fell". Until the 18th century many people still believed that Hekla was "The Gateway to Hell". There are still superstitions which exist to this day of witches gathering on Hekla for Easter.

Hekla sits astride the middle of the mid Atlantic spreading zone hot spot, called a rift – transform junction, where the south Iceland seismic zone meets the eastern volcanic zone. Hekla is a mixed type volcano being a mixed type of crater format and strato volcano. This is a very rare format found in very few of the word's volcanoes, one of these is Callaqui in Chile in South America.

As previously mentioned the primal fissure opening is called the Heklugja Fissure which opens upon its entire length in an eruption which is fed by swiftly flowing magma from a massive magmatic reservoir estimated to have a roof some 4 kms below the surface with a centre of the magma chamber 2.5 kms lower down in the earth's crust. The ash produced by Hekla is very high in its concentrates of fluorine, which is poisonous to humans and animals. Hekla's basaltic andesite lava generally has a $SiO2$ content of over 54% compared to the 45 – 50 % of other nearby transitional alkaline flood basalt eruptive lava's. Hekla is the only volcano to produce calc – alkaline lavas. Phenocrysts in Hekla's lava can contain plagioclase, pyroxene, titanomagnetite, olivine and apatite amongst other chemicals and crystalline elements extruded from the volcano.

When dormant Hekla is covered by snow and glaciers and ice caps, it is a fairly non seismic volcano with seismic activity only starting to occur between 30 – 80 minutes before an eruption, this is one of the reasons why Hekla and similar volcano's are so dangerous, a jet airliner flying fairly close to Southern Iceland at 35,000 feet would be perilously close to Hekla if it suddenly blew without warning. In 1982 a BA flight, which was flying over the Indonesian Islands in the dark, flew through a huge ash cloud at night without realising what was happening, apart from the plane's external surfaces glowing an eerie green phosphorescent colour, suddenly the engines stopped working one by one and the plane started a crash unpowered glide. The British pilots kept calm and didn't panic and kept on attempting to restart the jet engines, the plane was getting closer and closer to the sea and the pilot and co pilot must have thought their final moments had come, then amazingly one of the engines came back on line and powered up, then another, then another until all the jets engines were working and the captain pulled hard back on the stick and the jumbo jet raised its nose and started to regain height and the captain then headed for the nearest airfield in Indonesia to put the 747 down in. Eventually the plane landed and all on board survived the trauma. What had happened when the plane flew into the hidden (by nightfall) ash cloud, which incidentally does not show up on a plane's radar screens (this has now been rectified and modern jets have an ash particle detector on board now I am reliably informed) was that the heat of the jet engines turned the silica ash into glass which then started to cover all the parts of the engine with a layer of volcanic · glass, eventually this glass covered the fuel ducts into the jet engine causing the engine to cease working. The eerie green glow was the electrically charged particles of ash discharging against the aircrafts surfaces a little like St Elmo's Fire on board a ship.

The engines all cut off, the plane then dropped towards the ocean in its glide path and as it came out of the ash cloud the glass cooled quickly, cracked, was blown off by the slipstream and then the fuel could flow again and as the captain and engineer were constantly trying to restart the engines and were following prescribed aviation procedures and had not panicked which in effect saved the lives of everyone on board the plane, once the glass had gone the engines could be restarted and that saved the day for all those onboard the flight. If the plane had been flying at a lower altitude then there might not have been enough time to restart the engines, this incident woke up the CAA and other authorities to the very real threat of volcanic ash being ingested by aircraft engines, as well as liquefying glass covering engine and aircraft surfaces there was very bad scouring and gouging of the airplane from the friction effect of the plane hurtling through the ash cloud and all the small pieces of stone, ash, silica rubbing against the aircraft surfaces, this can cause major damage especially to the delicate parts inside a jet aircraft engine, similar to the effects a sand storm would have on a jet aircrafts air intakes.

Much work has now been done by scientists in trying to ascertain what is the safe level of of ash particles per square metre in a volcanic ash cloud which is acceptable to allow the aircraft to continue flying through the ash cloud, scientists have come up with a figure approximately 3 particles of ash per square metre, anything more than that could be harmful to engines and bodywork of an aircraft and could theoretically bring the plane down. The British Met Office now issue a regular ash cloud warning service which can be used by all airlines to monitor which are safe areas to fly in, in the event of an eruption similar to EJ in March – April 2010.

Hekla due to its turbulent past is constantly being monitored, scientists are checking for strain, deformation, tilt and other movements and seismic activity, if a magnitude 3 or more seismic shock is registered then you had better take cover as this indicates that Hekla is about to erupt.

Hekla's Eruption History (prehistoric and historic)

Going back in time to about 1000 – 1100 BCE into prehistory was the Hekla 3 eruption (H3), which ejected approximately 7.3 km3 (cubic kilometre's) of tephra into the atmosphere, making its volcanic explosivity index (VEI) at 5. This mega explosion would definitely have brought on a mini ice age weather pattern for the Northern Hemisphere for at least 3 – 5 years, in London we would be getting the sort of weather that Moscow normally receives, we will need to import some more vodka to keep up our spirits during the next mini ice age.

Evidence of this eruption has been found in Scottish peat bogs and in Ireland a study of dendrology (the study of dating tree rings) has shown that there was no noticeable growth of tree rings for nearly a decade, showing that the eruption had cooled down the Northern Hemisphere considerably and freezing cold winters and summers with little heat and possibly even snow during the summer months were the norm during these 10 years. It would be hard to imagine how normal everyday life in the UK and Europe would cope with this sudden shift in temperature of this magnitude, approximately 10 degrees colder in winters, it would be like moving London over a period of a few months into the middle of Northern Canada for 10 years, forget about the wrong sort of snow and the wrong sort of leaves, the tube trains would be frozen solid and we would have to learn to drive Sno Cats (special Antarctic vehicles with four sets of caterpillar tracks and diesel engines modified for arctic conditions) and skidoos (petrol powered ski sleds with tracks) and get our husky dog team out of the garage for the weekly shop to the Tesco igloo.

Many people have wondered what brought on a so called "dark ages" of early Greek civilizations at Mycenae and Anatolia as well as in the late bronze age across the Middle East in Israel, Lebanon, Egypt, Syria and other areas of emerging civilizations, there was suddenly across all these areas a complete collapse of central authority, towns emptied, villages seemed to have been deserted, all commerce for about 100 years seems to have dwindled. I would speculate that one, or a series of major volcanic eruptions either in Iceland, North America, Indonesia or elsewhere caused weather patterns to change abruptly, causing much harsher, colder or stormy weather to cause major problems to cultivated areas. If the crops failed year after year then civilised life would not have been able to survive and cities and towns would be deserted as the survivors went elsewhere to try and seek food and water.

Cities like Jericho and Troy seemed to have been abandoned at this time and only rebuilt many years later on a smaller scale, these enigmas will hopefully one day be proved to be caused by shifting weather patterns caused by large-scale volcanic activity elsewhere on the Earth.

Obviously about 1500 BCE Santorini (Thera) blew itself off the map and although the resulting tsunami's probably destroyed the Minoan Civilization on Crete and damaged other coastal areas of Turkey, Israel and Egypt its effects were not long lasting, however the likes of Hekla, Katla, Bardarbunga and the other huge Icelandic volcanoes as well as one of the many super volcanoes situated around the globe could have contributed to this late bronze age mega disaster. This book is mostly to do with enlightening the general public to the fact that we live on the edge of very serious dangers and although governments spend millions of pounds on environmental issues such as global warming and a comet or asteroid strike, perhaps we should be spending monies on some sort of early warning system for major volcanic activity, which to my mind poses one of the most dangerous threats to modern civilization and is the least known about of all the above threats.

The main other prehistoric Hekla eruptions are H5 in 5050 BCE, H – Sv in 3,900 BCE, H4 in 2310 + or – 20 BCE and as we have discussed H3 in 950 – 1100 BCE. H,3,4 and 5 produced massive amounts of rhyolitic ash and tephra covering 80 % of Iceland and was detected in the soils of Europe as far away as Scandinavia, Europe and The Orkneys. H3 and H4 produced the largest amount of tephra since the last Ice Age about 10,000 years ago.

1104 (H1) Eruption

Hekla had been relatively quiet for approximately 250 years when it suddenly and unexpectedly erupted in the latter part of 1104 ACE. This huge eruption covered literally over half of Iceland's 55,000 square kilometres with approximately 2.5 cubic kilometre's of rhyodacitic tephra. This was probably the second largest tephra eruption in recorded historical times with a VEI (volcanic explosivity index) of 5 like the previous H3 eruption. Many farms which lay in the path of the ejected tephra over 15 kms away such as the Pjorsardalur Valley, and Hrunamannaafrettur, which was 50 kms away as well as at Lake Hvitarvatn, which was 70 kms away, were abandoned due to the harsh conditions of tephra, lava bombs, lava and other volcanic damage done to the area. This was the main eruption that caused Hekla to become something of a cause celebre across Europe in the Middle Ages and the whole of Europe became aware of the "enfant terrible" that was Hekla.

I would imagine that if this sized eruption happened today then the consequences for Europe would be pretty dire, cancellation of all flights to Europe for many months, thick ash coming down like snow in many parts of Europe, major disruption of industry, travel, communications, and food distribution across Europe for months on end, we just have to hope that Hekla stays slumbering for a few more years and does not awaken with a bang as in the past. One of the most worrying things about Hekla is that we would get very little or no warning before it erupts as due to its particular geological formation and where it sits on the mid Atlantic Ridge, previously there has been very little or no warning of imminent eruptions.

1158, 1206 and 1222 Eruptions

A VEI of 4 eruption started on January 19th 1158 producing over 0.15 km3 of lava and 0.2 km3 of tephra. The 1206 and 1222 eruptions which first began on December 4th were much smaller eruptions VEI of 3 and 2 and distributed around 0.24 km3 of tephra mainly to the north east of Iceland, these were not nearly as catastrophic as H1 (the 1104 AD) eruption and did not cause any major migrations of farmers across the island.

Iceland unlike Indonesia and Hawaii and other tropical and sub tropical areas of volcanicity has a wealth of data from local annals, monks chronicles and folk tales which is why we now have a much better historical view of all the major eruptions since the 9th century ACE.

1300 – 1301 Eruptions

This series of eruptions was a VEI 4 and started on July 11th and lasted for a year. This was nearly as big as the 1104 eruption and again one of the biggest Iceland had seen in historical times. It covered 30,000 square kilometres of land with 0.31 km3 of tephra. Nearly 0.5 km3 of lava was produced by this eruption and due to this volcanic activity major damage was caused to the villages of Skagafjordur and Fljot leading to more than

520 deaths the following winter. There were high levels of SiO_2 of between 56 % and 64 % and many farmers in Iceland must have considered immigration or moving to a different part of the island as a result of these eruptions and their damage to livestock and property.

1341, 1389 and 1440 Eruptions

In 1341 a relatively small eruption of VEI 3 starting on May 19th caused small amounts of tephra and lava to go over areas to the west and southwest of Hekla leading to many cattle deaths in the vicinity probably from flourosis poisoning. Then the actual eruption fissure moved itself away from the volcano itself and into some wooded areas above Skard.

Skard and other nearby farms were destroyed by a huge lava flow that now forms an area today that is called Nordurhraun, about 0.3 kms of lava and some smaller amounts of tephra were produced by this eruption sequence. Further rumblings around 1440 may have occurred, however there is no definite evidence for this.

1510 and 1597 Eruptions

An eruption in 1510 occurred which was VEI 4, which is an extremely violent eruption. Volcanic bombs fell over 40 kms to the west at Vordufell. Tephra has also been found at Rangarvellir, Holt and Landeyjar.

In 1597 a VEI 4 eruption started on January 3rd and carried on for over 6 months. Over 0.15kms3 of tephra was deposited to the south east damaging Myrdalur.

1636 – 1637 and 1693 Eruptions

In 1636 although the VEI rating was low, about 3 on the scale, the eruption continued for over a year, damage to pasture's and death of some livestock occurred.

The 1693 eruption which lasted for about 7 months was one of the strongest and most powerful eruptions of Hekla for hundreds of years and was a VEI 4 level eruption. Approximately 60,000 cubic metres or 0.18 km3 was produced and lahars (volcanic produced floods of mud and water) and some tsunami were also recorded. Wind blown tephra was deposited in the north west of Iceland damaging farms and wooded areas in Thjorsardalur, Land, Hreppar and Biskupstungur. Ash from this eruption was detected as far afield as Scandinavia and the Orkney Islands. Much destruction was caused to local wildlife in Iceland as well.

1725, 1766 – 1768 Eruptions

A small eruption took place in 1725, possibly VEI 2 on April 2nd, producing some lava flows, these are technically not actually classed as Hekla eruptions although they do come from the same volcanic area and system as Hekla. We know this from analyzing the SiO2 content of the lava itself.

A much more serious and larger eruption occurred, starting on April 5th 1766 at 3.30am in the morning. A huge lava flow of 1.3 km3 covered 65 km2 of land and the explosivity index (VEI) was 4 on the scale. The amount of tephra produced was 0.24 km3 which is the third largest volume since recorded history commenced. Firstly a 2 – 4 cm layer of ash was laid over Austur – Hunavatnssysla and Skagafjordur causing the death of many fish and cattle. Some small farms and villages also suffered damage. Lava bombs of up to 0.5 m in size were thrown 15 – 20 kms away and some localised flooding resulted from the melting of snow and ice on Hekla's upper slopes.

1845 – 1846, 1878 and 1913 Eruptions.

Hekla had slumbered quietly for more than sixty years prior to 1845, when on 2nd September at 9 am of that year it suddenly, without any warning exploded. Below is an anonymous description of this eruption;

"After a violent storm on the night of 2nd September (1845) in that year, the surface of the ground in the Orkney Islands was found strewn with volcanic dust (tephra). There was thus conveyed to the inhabitants of Great Britain an intimation that Hecla (sic) had been again at work. Accordingly, tidings soon after arrived of a great eruption of the mountain. On the night of the 1st September, the dwellers in its neighbourhood were terrified by a fearful underground groaning, which continued till mid day on the 2nd. Then with a tremendous crash, there were formed in the sides of the cone two large openings, whence there gushed torrents of lava, which flowed down two gorges on the flanks of the mountain. The whole summit was enveloped in clouds of vapour and volcanic dust. The neighbouring rivers became so hot as to kill the fish, and the sheep fled in terror from the adjoining heaths, some being burnt

before they could escape. On the night of the 15th September, two new openings were formed – one on the eastern, and the other on the northern slope – from both of which lava was discharged for twenty two hours. It flowed for a distance of upwards of twenty miles, killing many cattle and destroying a large tract of pasturage.

Twelve miles from the crater, the lava – stream was between forty and fifty feet deep and nearly a mile in width. On the 12th of October a fresh torrent of lava burst forth, and heaped up another similar mass. The mountain continued in a state of activity up to April 1846: then it rested for a while and began again in the following month of October. Since then, however, it has enjoyed repose. The effects of these eruptions were disastrous. The whole island was strewn with volcanic ash, which, where they did not smother the grass outright, gave it a poisonous taint. The cattle that ate it were attacked by a murrain, of which great numbers died.

The ice and snow which had gathered about the mountain for a long period of time was wholly melted by the heat. Masses of pumice weighing nearly half a ton were thrown to a distance of between four and five miles"

—— *Anonymous 1872.*

This was a huge eruption and did not finish for a year until April 1846, about 20,000 cubic metres of tephra was produced in this VEI 4 blast. The tephra (0.17 km3) was distributed mainly to the south east of the island and immediately to the east of Hekla the tephra was 20 – 40 cms deep. Lava flows to the west and northwest of the island covered an area of 25 km2 with 0.63 km3 of lava. Many cattle and sheep died throughout the year and in the following year due to the large deposition of dark ash over the pastures of Iceland. Ash was carried by the prevailing winds to the Faroe Islands, Shetland and The Orkneys and some ash may have been blown as far as Scotland and Scandinavia, this eruption had it occurred today could well have closed the air routes to Europe for much of the year and caused major disruption to commerce, shipping, transport and food supplies to Europe and surrounding areas.

There were several small eruptions between February 27th 1878 and April 1878 and then on April 25th 1913 and May 18th 1913. These were not actually from Hekla itself but from fissure's about 10 kms east of Hekla. The first eruption produced 0.2 kms3 of lava from two parallel fissures covering 15.5 km2. The second caused large fissures at Mundafell and Lambafit which then produced about 3.8 km2 and 6.3 km2 of lava respectively.

1947 – 1948 Eruptions.

The 1947 eruptions started on March 29th 1947 and ended on April 21st 1948. This eruption was a serious VEI 4 on the explosivity index, it has been calculated that this was both the second greatest lava eruption of Hekla whilst Iceland was inhabited and the second greatest lava eruption in the world for the period 1900 – 1970. A total lava volume of 0.8 km3 was produced with

0.21 km3 of tephra. The actual height of Hekla was 1447 m before the blast and afterwards it increased to a new height of 1503 metres before dropping to 1491 metres today.

The First Eruption of 1947

This eruption occurred over 100 years since the previous eruption of Hekla itself, which was the longest dormant period since 1104. It has been deduced by the scientific community that the longer Hekla lies dormant, the bigger the subsequent eruption tends to be, so in our situation we would be happy to see fairly small scale localised eruptions every 10 years or so rather than these monster sized eruptions that occur every 100 years or so.

Prior to the eruption Hekla had been visible, however nothing untoward was noticed about it. The actual eruption occurred without warning on 6.41 am with a huge roaring noise, the noise of later eruptions could be heard throughout Iceland. An earthquake at 6.50 am measured 6 on the Mercalli Intensity Scale and increased the level of eruptions until the eruptions were covering a 4 km fissure on the ridge. For the first few hours of the eruption it was plinian, later on it turned into a vesuvian style eruption. The ash cloud had risen to a height of 30 kms by 7.08 am, the wind then carried it southwards towards our old friend Eyjafjallajokull (EJ to you and me) turning the glacial ice and snow black. Pumice and ash first landed on Fljotshlid at 7.10am and tephra and ash continued to fall making a layer 3 – 10 cms deep. A lava bomb landed 32 kms from Hekla and was 0.5 m across and weighed 20 kg. Between Vatnafjoll and Hekla a layer of tephra up to a metre thick built up including many large lava bombs. Bombs with an area of 50 m2 were dropped onto the slopes of Hekla up to 1 km distance from the ridge. 51 hours after the eruption had started ash started to fall on Helsinki in Finland having travelled 2860 kms in this short period of time.

If this eruption had happened today and the wind had been blowing to the south east of Iceland, much of Europe would be in a no fly zone within about 20 hours of the eruption with no prior warnings or notice given, as stated previously: today's high tech, computer enhanced age, our susceptibility to disruption from this sort of problem would be much more serious. In 1947 the level of transatlantic travel by air was negligible and most freight was carried by ship so the problems then would be much more visible and immediate nowadays. Much of the damage to our way of life would ultimately depend on what time of year the eruption occurred as the prevailing winds do normally blow west to east from the USA to Europe, as can be noticed if you are an airline passenger, due to the effects of the jet stream which is a fast flowing stream of air above the Atlantic at approximately New York to London Latitudes, this means that going to the USA takes about 7.5 hours, however coming back only takes 6 hours. If the wind direction is SE towards the UK and EU then we will be right in the ash cloud track and would feel the full effects within 72 hours, however if there was a ridge of high pressure over Iceland and the winds were not directly blowing west to east, then the ash fallout might travel over the Arctic or Canada, we just might get lucky in those circumstances. However for the purposes of being prepared for the worst we cannot rely on the vagaries of weather systems and should always prepare for the worst outcome whilst hoping for the best outcome.

The amount of tephra being produced in the first 30 mins of the eruption was 75,000 m3 s -1 dropping to 22,000 m3 s -1 for the next half hour. The first cycle produced 0.18 km 3 of tephra which covered 3,130 km 2 of land and sea. 98 farms were damaged by the eruption. A huge volunteer effort was mobilized to clear the tephra, around 1000 man days by the end of July. The eruption produced around 3 Ml of water (snow melt and directly from the fissure) which caused the Ytri Ranga River to burst its banks and flood the surrounding areas.

3,500 m 3 s -1 of lava was produced in the first 20 hours of the eruption which divided into various fissure branches and covered 12 – 15 kms. On day 2, 8 distinct eruption columns were seen by observers. A huge lake of lava was formed which was 860 m wide was called Hraungigur, which produced a constant flow of lava down the mountain slopes. Another crater named the shoulder crater (Axlargigur) produced a column of smoke every 10 seconds together with loud explosions that created visible compression waves in the smoking vents. By the fourth and sixth days the eruptions were generally dying down and subsiding and only the shoulder and summit craters were erupting explosively.

Final Stages in the eruptions

The main explosive eruptions increased in size from 9 – 12th April 1947 and from 28th April reduced again. On

May 3rd the volcano stopped erupting lava in sudden explosions from its craters and changed to continually ejecting tephra and ash for long periods, until June when these ejections reduced in size. Sandy textured tephra and ash fell over Iceland in May and June sometimes making it seem like it was night time near the volcano. The tephra caused fluorine poisoning of grazing sheep making them unable to walk. In the winter of 1947 more volcanic craters were formed building up cone like structures.

Explosive volcanic activity had now ceased by August 1947. The lava sometimes ran through lava tubes (tunnels). The lava front had a height of up to 15 metres. On June 15th and 16th a branch of lava flowed to the south of Melfell and it travelled over 1 km in 30 hours before stopping on June 21st 7.8 kms from the lava crater. The longest lava stream produced was 8 kms long and stopped in Storaskogsbotnar. A scientist who was attempting to film the lava stream on November 2nd was hit by a block of lava and unfortunately later died of his injuries.

The lava flow finally stopped after 13 months on April 21st 1948 having covered 40 km 2 of land with a maximum depth of 100 m. The lava beds produced were mostly A a lava type with Pahoehoe and lava a budella areas. In April and May 1948 carbon dioxide (CO_2) emitted from cracks in the ground pooled in hollows near Hekla killing 15 sheep and some wild animals and birds. In total 24,000 tonnes of CO_2 was emitted, this would not have helped in controlling greenhouse gas emissions and may have helped to cause some local anomalies. Farmers dug dykes to drain these hollows of the CO_2 and by the end of 1948 these CO_2 emissions had ceased altogether.

In 1947 – 1948 there were some severe weather anomalies which the Hekla eruption could have played a part, if it can be shown that certain gases were leaking from Hekla before March 1947, or other matters which could affect weather patterns, could have affected the cycle of weather in that year then that could in part explain why 1947 – 1948 was one of the coldest years in the UK and Europe on record. I feel that further research into volcanic eruptions and local and international weather conditions need further work as I feel that there is a definite correlation between volcanoes and their associated systems and weather patterns and temperature gradients.

1970 Eruptions.

The 1970 eruption of Hekla started at precisely 9.23 pm on May 5th 1970 and lasted until July 5th. It had a VEI of 3 and produced 0.2 km 3 of lava covering an area of 18.5 km2 and 6.6 × 10 7 m 3 of tephra deposited over an area of 40,000 km2, mainly to the north west of Hekla.

Initial Eruptions.

A large amount of melted snow was noticed on the flanks of Hekla before the eruption, indicating that the volcano was heating up ready to explode yet again. Earthquakes measuring level 4 on the Richter scale were measured in the run up to the eruption. The eruption started slowly with only little ejecta at 9.23pm the first ash fell on Burfell power station 15 kms away at 9.35pm starting an evacuation of the facility and the surrounding areas. The eruptions seemed to have started in two places simultaneously, one in the shoulders crater and another below the lava crater. At 10.30pm a crater at 780 metres was producing a lava column which reached an altitude of 1000 metres. During that night a 700 metre lava fountain was thrown up from the main crater. A 500 metre long fissure starting below the lava crater opened and lava fountains and other lava flows occurred. Later on that night more fissures opened and further lava fountains and lava flows appeared. Next morning the lava flows had extended to 7.5 kms, the flow rate was an impressive 1500 m 3 s -1.

Tephra was being ejected at the rate of 10,000 m3 s -1, the ash cloud from the eruption had now risen to 53,000 feet (16,154 m) which caused a loud lightning storm to occur. The tephra was blown northwards by the prevailing winds causing the sky to turn black in areas. 190 kms away in Blonduos, tephra fell from midnight until 2 am and ash fell on a trawler 330 kms away at 2am.

Later Eruptions

By 5.30 am on May 6th the lava flow was 4 kms long. Lava bombs were noticed on the slopes of Hekla, one bomb had an area of 6 m2 and probably weighed in at 12 tons. Xenoliths formed about 2 % of the eruptive material, the lavas were based on basalt, andesite, ignimbrite and sedimentary rocks types.

The eruption increased by May 12th at Skjolkviar with columns of steam rising to 2,500 metres. The eruptions then slowed down until activity ceased on May 20th. By then the lava field covered an area of 5.8 km2. Later, on May 20th, activity started again with further fissures opening and large amounts of lava fountains pouring out lava and tephra, lava flows increased and the height of the ridge rose by 100 m. However all activity then died down and by July 5th all activity had ceased again.

Effects

During many of the Hekla eruptions fluorine is produced and this sticks to the tephra, grains of tephra can have a fluorine content of 350 ppm (parts per metre) and fluorine poisoning can start in sheep with a diet of fluorine covered grass at a level of 25 ppm. At 250 ppm death can occur within a few days of ingestion. In 1783, when the great Laki eruptions occurred 79 % of the Icelandic sheep stock were killed by fluorine poisoning which in turn meant that about 25 % of the Icelandic population starved to death subsequently. After the 1970 eruptions although the fluorine content was only 0.2 % this still was enough to contaminate the grass, therefore 450 farms and 95,000 sheep were at risk. Some farmers kept their flocks of sheep indoors and let them eat hay, however many farmers did not have this option and it is estimated that fluorine poisoning killed 1500 ewes, 6000 lambs and some horses in the days and weeks after the eruption.

1980 and 1981 Eruptions

This VEI 3 eruption started at 13.28 on August 17th 1980 and lasted until August 20th 1980, This was a mixed eruption with lava and tephra production and a large steam column preceding this eruption. Fissure's opened along a 7 km length of the ridge area. Lava covered 22km 2 within 24 hours of the eruption starting. This eruption was highly unusual in that it was the shortest eruption since 1104, previous eruptions have lasted several months or nearly a year on occasion, this eruption only lasted 3 days.

The 1981 eruption was also very short lived, it began on 3am on April 9th 1981 with a VEI of 2 and produced a small amount of lava and it threw an eruption cloud 6.6 kms high into the atmosphere and formed a new crater on the summit of Hekla. The lava flows extended to a max of 4.5 kms from the volcano, the eruptions ceased on April 16th, scientists feel that this eruptive cycle is a continuation of the 1980 eruption sequence.

1991 Eruption

On January 17th 1991 a VEI 3 eruption occurred, finishing on March 11th 1991. This produced 0.15 kms of lava and also an amount of tephra. This eruption was preceded by some sulphurous smells and earthquakes. The eruption started as a plinian eruption producing an ash cloud which rose to 11.5 kms within 10 minutes, which had travelled 200 km north east to Iceland's coast within 3 hours.

Then andesite lava started flowing, eventually covering an area of 23 km2 to a depth of 6 – 7 metres. Fissures erupted with lava fountains as per previous eruptions, these fountains reached 300 metres in height. By the 2nd day of the eruption activity had ceased in all but one fissure, some lava flows occurred, however this moved fairly slowly and had a low viscosity.

2000 eruption

There was a small eruption in 2000, it started at 18.18 on February 26th 2000 and carried on until March 8th. It was a VEI 3 eruption producing a lava volume of 0.19 km 3, some tephra was also produced. The eruption went through four cycles, 1. early eruptive cycle, 2. eruption of fire fountains, 3. spurts of strombolian eruptions. 4. ejection of lava streams.

The eruptions were at their maximum in the first hour and by the first night the eruptive fissure vent on Hekla had opened to a length of between 6 – 7 kms. Steam rose in a column to almost 15 kms in height and ash went as far as Grimsey. In the middle of the eruption a NASA DC -8 jet plane accidently flew through the volcanic ash plume with all its measuring instruments switched on and the resultant cascade of data of an early eruptive cycle has proved very useful to the volcanologists monitoring volcanic eruptions in Iceland and elsewhere in the world.

Until recently it had been thought that Hekla was not the typical type of volcano that produced the extremely dangerous and ferocious pyroclastic flow, this is a phenomenon that occurs when some of the erupting ash cloud falls back onto itself and then travels at great speed, sometimes as much as 100 mph down the sides of the volcano engulfing and burning and destroying anything in its path. In Pompeii in 79 ACE, this is what finally destroyed the roman city, engulfing its inhabitants in extremely hot ash, stones, and pumice which flew down the sides of Vesuvius after part of the ash cloud had imploded upon itself. These flows are quite common in certain volcanoes including the recent eruptions in Montserrat and some in the Far East.

Dr Armann Hoskuldsson from the Norvol Institute in Reykjavik found some traces of pyroclastic flows measuring 5 kms in length going down the side of the mountain. This will mean that a reappraisal of volcanic eruptions of the basic rock type will have now to be undertaken, this sort of volcano up to now was thought not to create any of these pyroclastic flows. This new discovery could mean that the authorities in Iceland try and keep the public and tourists whom continually tend to rush to any new eruption, away from any potentially dangerous areas in case they are at risk due to these dangerous flows.

My current opinion

In the 20th Century and the first 10 years of the 21st Century Hekla has already erupted 7 times so far, that averages one eruption every 16 years approximately. The 1947 – 48 eruptions were of a similar explosive energy to the 1846 – 1847 eruptions, it would seem therefore that although Hekla erupts fairly regularly, that about every hundred years or so much bigger and more dangerous eruptions occur.

Without any doubt in my mind Hekla is a massive risk to 21st Century living, our infrastructure would be severely tested by a huge eruption of Hekla, not just the fact that all flights across the Atlantic could be cancelled for months on end, it would, with its poisonous ash cause major food production problems to many farmers across Europe and the knock on effect of poisonous ash, no air traffic, possible disruption to basic utilities and travel, would cause great suffering and massive inconvenience to our way of life.

Knowing human nature, if the population of Europe heard on the news that a major eruption of Hekla had started and a massive semi poisonous cloud of ash was heading for Europe I would imagine that all the supermarkets, corner convenience stores and other shops would be empty of food and supplies within three hours and that huge queues would start outside of all the petrol stations as people panicked about getting food, water and petrol supplies organised before the ash cloud arrived. This very fact would probably contribute more to the chaos than anything else, as then governments and suppliers and distributors would be on "the back foot" as the distribution network was not designed to cope with these sorts of emergencies. Suppliers would then be called to release large amounts of further stock so that shelves could be replenished, however I then feel that whenever further stocks arrived that the population would still be clamouring for more supplies and the shelves would empty as quickly as they were filled. This cycle would fairly quickly break the ability of suppliers and distributors to actually keep the shops stocked up and then the authorities may be brought in to ensure that any food and supply distribution was fair and equitable. Unfortunately, human nature means that people would attempt to hoard, fearing the worst, they would want to build up their supplies in case there was a further disruption of supplies, this very act of hoarding would actually possibly cause the problem that everyone would be fearing, a breakdown in the supply and distribution chain.

All this is without any actual initial effect from the erupting volcano, if the volcanic eruption was really severe the whole situation would be further compounded by lorry drivers not being able to get to depots, docks not being able to get supplies to central distribution areas and so on, leading eventually to the Government organising military conveys and a general emergency distribution regime.

For those sensible citizens who would have followed my earlier advice and stocked up, all the above problems would not affect them, the main issues I feel would be that other friends and families and neighbours would be going hungry or would be in need of help and then the decision of whom to help and how much to help would come into play. Obviously if you gave all your food and supplies away to every hungry soul then you and your family would go hungry and be totally dependent on the Government's distribution scheme, which might be a very basic bread, emergency rations and water scheme, depending on how efficient the scheme was. We shall again discuss these social and important issues later in the book.

Hekla is obviously a contender for being very near the top of the list of repeat offenders whom we all have to watch extremely closely, I am sure that the Icelandic authorities are closely monitoring Hekla and similar volcanoes in case of eruption signs. Our biggest problem with Hekla, Laki and some of the other strato volcanoes is that sometimes there are no fore warnings of imminent eruptions that are about to occur due to the nature of the Icelandic volcanoes, this is a significant problem as if Hekla was to erupt today, within about 10 minutes you could have a 15 km ash cloud disrupting aviation and causing all sorts of problems, especially if there were strong winds blowing from the N W to the S E, we could be on the receiving end of ash deposits within 48 hours the eruption starting.

Another problem with Hekla is the longer it slumbers the bigger the eventual eruption tends to be, our only consolation is that Hekla has let off steam, so to speak 7 times in the last 110 years so it is doubtful although not impossible whether an 1100 ACE or similar mega eruption will occur in the next few years time. I feel that a lot might depend on how much melting ice due to global warming takes the weight off some of these volcanic systems as in releasing the cork in a champagne bottle and as the magmatic chambers fill up and the earth swells it may be that several of these volcanoes blow at the same time or trigger each other off as in EJ sometimes sets of Katla over past centuries.

I will of course keep a watchful eye over any data I receive from the various Icelandic Institutes and post my observations weekly (or daily if applicable) on www.icelandicvolcanoes.co.uk

In my opinion Hekla could be definitely the one volcano that could cause us in Western Europe major problems if and when it next erupts.

Danger Level : 9 (out of 10)

Eruptability Level : 65 % probability of an eruption in the next 10 years.

10. Hengill

By kind permission of Mr Oddur Sigurdsson jardfraedingur (geologist) Icelandic Meteorological Office, Reykjavik, Iceland

"An aerial photo of Hengill situated to the South West of Iceland, it covers an area of 100 km2. The volcano is still active and is an important source of geothermal energy supplying Reykjavik and other towns with power and hot water, there are numerous hot springs and fumaroles around the caldera area."

Style: stratovolcano
Location: Situated in S W Iceland, to the south of Pingvellir.
Height: 803 m (2,635 ft)
Last Erupted: 90 ACE

The Hengill volcano is an active volcano which covers an area of 100 km2. Hengill is definitely still in play, i.e. still an active volcano and produces numerous hot springs and fumaroles. The last eruption of any note occurred about 2000 years ago.

Hengill is important to the Icelandic economy as it produces much of Iceland's free geothermal energy. The hot water vents are captured at the Nesjavellir Power Station (near the western side of Lake Pingvallavatn) and the Hellisheioi Power Station which is situated 11 kms S W of Nesjavellir). Both power stations are operated by Orkuveita Reykjavikur (Reykjavik Energy).

The small town of Hverageroi has a multitude of hot springs which are also linked to Hengill.

My current Opinion

Hengill appears to be a minimal threat to Europe and the UK currently, as it has not erupted for over two thousand years. Of course that does not mean that it won't erupt at some time in the future and if it is at all like Hekla then it will make an almighty bang when it does eventually erupt.

The fact that there are stable geothermal springs and vents suggests though that it is slumbering peacefully for the moment and supplying the capital, Reykjavik with loads of free hot water and power for all the inhabitants, maybe we in the UK should contemplate moving to Iceland with its free energy, pretty young Icelandic maidens and outdoor activities abounding, also if we are upwind of erupting volcanoes then we wouldn't be that affected if we were living in Iceland. However a note of caution should be made here, eruptions over the ages have forced many Icelandic folk to migrate abroad or starved them to death and destroyed their livelihood in that thousands of livestock, cattle and sheep have been killed by the numerous eruptions over the ages.

Danger Level: 1 (out of 10)

Eruptability Level: 5 % probability of eruption in the next 10 years.

11. Herdubreid

By kind permission of Mr Oddur Sigurdsson jardfraedingur (geologist) Icelandic Meteorological Office, Reykjavik, Iceland

"An aerial view of the area containing the Herdubreid volcano, the actual volcano is a Tuya style volcano, carved by glaciers in prehistory. It is situated in the middle of the Odadahraun desert and close to Askja volcano. It originally sat under the Vatnajokull glacier which was much larger during the last ice age."

271

Style: Tuya type volcano (Tuya means a flat topped, steep sided volcano which has erupted through a glacier or ice sheet)

Location: Situated in Eastern Iceland in the Odadahraun desert and near to the Askja Volcano.

Height: 1,682 m (5,518 ft)

Last Erupted: In the Pleistocene era (2,588,000 BCE – 12,000 BCE)

Herdubreid Volcano is located in the midst of the Odadahraun Desert which is a large lava field originating from eruptions of Trolladyngja Volcano. The volcano originally was located beneath the Vatnajokull glacier which covered a much larger area during the last ice age.

Due to the mountains inaccessibility and steep sides this mountain was not actually climbed until 1908.

Near the mountain lies an oasis called Hedubreidarlindir which includes a camping ground and hiking trails. In previous centuries this oasis was inhabited by outcasts and criminals whom were excluded from Icelandic society, including the famous Icelandic outlaw Fjalla – Eyvindur who lived there during the winter of 1774 – 1775, who sounds much like the English Highwayman Dick Turpin.

My current opinion

Herdubreid sounds like a dormant or slumbering volcano to me and I don't feel that there are any possibilities of this one erupting in the foreseeable future.

Danger Level: 1 (out of 10)

Eruptability Level: 5 % probability of an eruption in the next 10 years.

12. Hofsjokull

By kind permission of Mr Oddur Sigurdsson jardfraedingur (geologist) Icelandic Meteorological Office, Reykjavik, Iceland
"An aerial view of Hofsjokull Glacier, which is the third largest glacier in Iceland. Hofsjokull is the largest active volcano in Iceland today, it covers an area of 925 km2 and is a subglacial shield type of volcano with caldera."

Style: Subglacial shield type of volcano
Location: N W of the Highlands of Iceland and North of the mountains of Kerlingarfjoll.
Height : 1,765 m (at summit)
Last Erupted: unknown (possibly some time in the Holocene era)

Hofsjokull is the largest active volcano in Iceland, situated between the two largest glaciers in Iceland. It covers an area of 925 km2 and is designated as a sub glacial volcano of the shield type which means it erupts over a wide area and is flatter than the normal conical style of strato volcano. Hofsjokull lies along an east – west area which connects the two principal rift faulting zones in Iceland. It forms a bridge between the Reyjkjanes – Langjokull rift on the west, which ends at Langjokull and the Eastern zone which extends N E across East Central Iceland. The main caldera is roughly 7 × 11 km and lies beneath the western part of the massive Hofsjokull icecap. Flank volcanic fissure volcanoes to the north and east of the icecap have produced some large flows of flood basalts during the Holocene era.

Kerlingarfjoll is another volcanic area which is part of the Hofsjokull volcanic system, it is dissected by glaciers and largely dates from the Pleistocene era, and is based S W of the main icecap. Its lava domes produce numerous hot springs which occupy two calderas at the centre of the 7 × 5 km wide volcanic area. Many vigorous fumaroles' venting smoke and steam are concentrated at the centre of the complex making this area one of the most geothermally active areas in Iceland today.

My current opinion

Although Hofsjokull appears to be the largest (in size) of all the active volcanoes on Iceland today, from its history and layout I doubt if this massive slumbering monster of a volcano will erupt in the near future.

As we mentioned before, if the geothermal activity is regular and no seismic activity has been reported then I would not put this volcano on my top ten volcanoes to watch carefully at the current time of writing.

Danger Level : 2 (out of 10)

Eruptability Level: 15 % probability of an eruption in the next 10 years.

13. Hverfjall (also known as Hverfell)

By kind permission of Mr Oddur Sigurdsson jardfraedingur (geologist) Icelandic Meteorological Office, Reykjavik, Iceland

"A view of Hverfjall volcano which is a tephra cone or tuff ring style volcano situated in Northern Iceland to the east of Myvatn. It last erupted in 2500 BCE and is today fairly quiet."

Style: Tephra cone or tuff ring style volcano
Location: Located in Northern Iceland to the east of Myvatn
Height: 420 m (1,378 ft)
Last Erupted: 2,500 BCE

Hverfjall last erupted in 2,500 BCE in the southern part of the Krafla fissure swarm, the deep crater is approximately 1 km in diameter. Tephra from Hverfjall has been deposited all over the Lake Myvatn area. A landslide apparently occurred in the south part of the crater during the last eruption which accounts for the distortion of the roundish shape of the volcano. During historical times, a lava flow from Svortuborgir to the south of the Namafjall Mountain, around Mount Hverfjall, which was almost totally immersed by the lava flow. Simultaneously another eruption occurred in the sides of the valley of Hlidardalur.

My current opinion

Hverfjall seems currently to be dormant, I feel that in the short term this is not one of the more dangerous of the Icelandic volcanoes. I truly hope that I am not proved wrong and that one of these quietly slumbering dangerous giants does not awake and cause much mischief and mayhem across Europe and the Northern Hemisphere.

Danger Level: 2 (out of 10)

Eruptability Level: 15 % probability of an eruption in the next 10 years.

14. Katla (\Narning Levell risk)

"A rare photo ofKatla majestically erupting in 1918, Katla is one of the most regularly erupting powelful volcanoes on Iceland and some of the tephra erupted in prehistoric times can be found over much of Europe. The 1918 eruption set off some massive flooding of the surrounding areas, these volcanic based floods are called jokulhlaup."

"A more recent photo of the mighty Katla with an unsuspecting horse grazing peacefully in the foreground. One of the main worries about Katla is that when Eyjafjallajokull erupts, normally within a short period of time Katla erupts as well, there are strong indications that there is some magmatic link between the two volcanoes."

By kind permission of Mr Oddur Sigurdsson jardfraedingur (geologist) Icelandic Meteorological Office, Reykjavik, Iceland
"A recent aerial view of Katla, showing the massive size of the surrounding ice cap."

Style: Subglacial fissure style volcano
Location: Situated in Southern Iceland, located to the North of Vik i Myrdal and the east of the glacier, Eyjafjallajokull (yes our old friend EJ again, it is a glacier and a volcano, very confusing)
Height: 1,512 m
Last Erupted: 1918

Katla, like Hekla and Bardarbunga and Grimsvotn and the other dangerous volcanoes is one of the volcanoes that need constant watching and monitoring. It is a highly explosive and historically proven, extremely dangerous volcano and as I mentioned earlier, when its good friend EJ erupts, normally Katla is not far behind. I am sure there are quite a few slightly sweating volcanologists in the Icelandic Institute of Volcanology who are nervously watching seismic and other data from Katla to see if it is waking up from its long sleep since 1918. If you suffer from high blood pressure or are of a nervous disposition you might not want to read the next few paragraphs.

The caldera of the volcano has a diameter of 10 kms (6 miles) and is covered with up to 2,300 feet of ice at the deepest areas. Katla normally erupts every 40 – 80 years, so it has now been 90 years since the last major eruption, so I would venture that we are now all living on borrowed time and I would nip out to Asda's now with your shopping trolley, just in case. The flood water discharge at the height of the 1755 eruption has been calculated to be 200,000 – 400,000 m 3/s (7.1 – 14.1 million cu ft/sec), comparable to the combined average flows of the Amazon, Mississippi, Nile and Yangtze rivers (approx 266,000 m3/s or 9.4 million cu ft/sec).

It is surmised that Katla was the cause of the Vedde Ash (more than 6 to 7 cubic kms or 1,4 to 1,7 cubic miles) of tephra dated to 10,600 BCE discovered at a number of areas including Norway, Scotland, and many coastal areas around Scandinavia and Northern parts of Europe, this seems to be one of the largest eruptions worldwide of the late ice age era. There have been approx 16 eruptions documented in the historic era of Iceland. The latest eruption occurred in 1918, although there may have been some minor eruptions that could not break through the icecap in 1955 and as recently as 1999. The huge 1918 eruption actually extended the southern coast of Iceland by 5 kms due to lahars (mud and rock slides) bringing earth and rock deposits down towards the coastline. The most worrying matter in all the above data is that the current slumber is the longest sleep since historical records began.

An eruption in 934 ACE produced a massive 18 cubic kms lava flows, one of the world's largest lava flows in the Holocene era (era since the end of the last ice age about 10,000 years ago). Katla has been the cause of frequent subglacial basaltic explosive eruptions that have been among the largest tephra producing volcanoes in Iceland during historical times and has also produced numerous dacitic explosive eruptions during the Holocene era.

It should be noted that before the 1974 Hringvegur (Iceland's Ring Road) was built travellers were scared to cross the plains in front of Katla due to the frequent Jokulhlaup (glacial volcanic flash floods) and also the very deep river crossings were treacherous. After the 1918 eruptions the floods coming out of the glacier were devastatingly large and dangerous.

Katla has been showing signs of waking up recently, which is especially worrying for us over here in Europe, this started in 1999 and geologists are monitoring this giant 24 hours a day. Since our old chum EJ erupted in March 2010, Note, EJ is also sometimes called Gudnasteinn (let's call it EJ aka GS) just to make matters more confusing. It is stated that the eruption of the long dormant EJ in March and April 2010 prompted major fears amongst some physicists that this could easily have triggered an eruption of the much larger and much more dangerous

Katla. It is thought that deep inside the Earth's crust or mantle that there could be subterranean passages or magmatic chambers which could be linking the two volcanic systems to each other so that when EJ blows or vice versa then Katla would also erupt. This theory has some foundation in truth as in the past 1000 years, all 3 known eruptions by EJ have triggered subsequent Katla eruptions. Following the 2010 EJ eruption on 20th April 2010 Icelandic President Olafur Grimsson said, "the time for Katla to erupt is coming close.... we [Iceland] have prepared.....it is high time for European Governments and airline authorities all over Europe and the world to start planning for the eventual Katla eruption." Please note Mr Cameron, you were warned and you can't blame Gordon Brown for this one.

My current opinion

Buy some strong ear plugs if you live in Aberdeen and Glasgow and surrounding areas of Scotland, as for sure when big K blows it will be front page news across the globe and we will be wading through tephra on Oxford Street in Central London on our way to work, if we haven't choked to death on fluorine gas or some other sulphurous brew. Let's hope it doesn't blow up before Xmas, we wouldn't want the shopkeepers and TV advertising revenue's to suffer and where would you go to buy a turkey if they had all suffocated in ash and gas. Also I did want to hear about the 40 % cut in Government spending before such polices get washed away by old Katla and a new realism about surviving and protecting the populations of the UK and Europe takes hold in the Governments of Europe.

One of the most worrying things I have researched is the longer Katla doesn't blow, the bigger it will blow when it does go up. It's as if, somehow the pressure from beneath the earth's crust pushes and pushes more magma into these underground chambers and then like a champagne cork popping it suddenly releases all that energy, and unfortunately unless someone can devise a way to tow the UK down to the Caribbean, we are more or less in the direct firing line of Katla (depending on wind directions and speed of course). Another trigger mechanism allowing Katla and other volcanoes to erupt fairly soon is the continuing global warming which is thinning the ice caps and glaciers of Iceland allowing the magma and pressure's from under the Earth to start pressurising these volcanoes again. So the combination of effects could result in a mega explosion fairly soon from one of the Icelandic volcanoes.

All one can do is to read my book and act on its contents and hope that perhaps some small eruptions under the icecap can slowly release the pressure of this "pressure cooker" of a volcano. I do feel that this volcano is "a clear and present danger" for life in Europe, this is an epoch changing volcano and could nearly be called a baby super volcano.

Danger Level: 9 (out of 10)

Eruptability Level: 85 % probability of an eruption within 5 – 10 years.

15. Kerlingarfjoll

"An aerial shot of the volcanic mountain range of Kerlingarfjoll which is situated in the Highlands of Iceland near the Kjolur highland road. There are still many hot springs in the area showing evidence of previous volcanic activity."

Style: stratovolcano of Tuya style
Location: Situated in the highlands of Central Iceland near the Kojolur highland road.
Height: 1,477 m (4,846 ft)
Last Erupted: Pleistocene era

Kerlingarfjoll is a volcanic mountain range in Central Iceland. Current evidence for these ancient tuya style volcanoes is evidenced by the numerous hot springs and small rivers in the area, they all form part of a large volcanic system of 100 km2 (38.6 sq miles). The earth is noted for its shimmering red colour and this is because the earth is made up from volcanic rhyolitic stones which in fact the mountains are mainly composed of. Minerals that have emerged from the local hot springs and fumaroles also colour the ground yellow red and green.

My current opinion

Low risk, an ancient volcanic system not much to worry about here.

Danger Level: 1 (out of 10)

Eruptability Level: 5 % probability of an eruption within the next 10 years.

16. Kolbeinsey

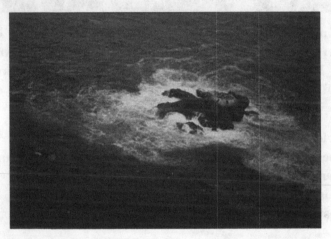

By kind permission of Mr Oddur Sigurdsson jardfraedingur (geologist) Icelandic Meteorological Office, Reykjavik, Iceland

Style: Basaltic submarine volcano (emerged as an island in 1372 approx)
Location: A small island 105 kms off the Northern coast of Iceland

Height: 8 metres above sea level
Last Erupted: Started erupting in the late Pleistocene era about 11,800 years ago. Last erupted approx 1372 ACE

Kolbeinsey is a small volcanic island approx 105 kms North of Iceland, it lies within the Arctic Circle and is one of the only parts of the mid Atlantic ridge which has emerged from the sea in this part of the world. It is basically a basaltic platform devoid of any vegetation and is being eroded by wave action at an alarming rate, it is expected to disappear from the surface of the sea in 2020.

In 1616 ACE its area was measured to be 700 metres (2,300 ft) from north to south and 100 metres (330 ft) east to west, by 1903 it had been reduced by wave action to about half that size. In August 1985 the size was measured to be 39 sq metres (128 ft) across. In 2001 it was measured again and found to have reduced to just 90 sq metres (970 sq ft) which equates to a circle 7.5 m (25 ft) in diameter, or the size of an average house. The island is only 8 metres high.

Helicopters had been able to land on Kolbeinsey as a concrete helipad had been laid out in 1989. On 9th March 2006 the helipad had been destroyed when a large piece of rock had become separated from the island, it is now thought that the island might disappear much more quickly due to erosional action of the sea.

My current opinion

No threat to anyone on Iceland let alone Europe.

Danger Level: 1 (out of 10)

Eruptability Level: 5 % probability of an eruption in the next 10 years.

17. Krafla

By kind permission of Mr Oddur Sigurdsson jardfraedingur (geologist) Icelandic Meteorological Office, Reykjavik, Iceland

"An aerial view of Krafla volcano situated in the North of Iceland in the Myvatn region. Krafla includes one of the best known Viti craters in Iceland. The Viti crater holds a green coloured lake in it. It is estimated that molten magma lies only 2.1 kms beneath the surface at this point in Iceland, Icelanders have harnessed geothermal energy from this region."

Style: A fissure style volcanic system
Location: Situated in the North of Iceland in the Myvatn region
Height: 650 m (2,133 ft)
Last Erupted: 1984

Krafla sits on one of the main N E to S W Fissure zones which are part of the Mid Atlantic Ridge spreading zone of volcanic and tectonic activity. The caldera is 10 km wide with a 90 kms long fissure zone. There have been over 29 reported eruptions in recorded historical times.

Krafla includes one of the two best – known Viti craters in Iceland (The other is in Askja Volcano). The Icelandic word Viti means "hell". In earlier times the Icelandic and Scandinavian people thought hell to be located under volcanoes. The crater Viti has a green lake inside it.

The Krafla area includes Namafjall, a geothermal area with many boiling and bubbling mud pools and hot steaming fumaroles. "The Myvatn Fires" occurred between 1724 – 1729 when many of the longitudinal fissure vents appeared to open up. The lava fountains could be seen in the south of the island and a lava flow destroyed 3 farms near the village of Reykjahlid, no one was injured in that event.

Between 1975 and 1984 there was some further volcanic activity inside the Krafla volcano. The sequence involved 9 volcanic eruptions and fifteen deformation uplifts and subsequent subsidence events. These activities interrupted some of the Krafla drill fields. During these volcanic events a huge subterranean magma chamber was discovered by the scientists, this was discovered by using the seismic data recorded during the tectonic upheavals.

Since 1977 Krafla has been used as one of Iceland's major geothermal heat centres. It powers a 60 MWe power station nearby. In 2006 a further survey indicated that there were very high temperatures at depths between 3 and 5 kms underground. These results caused the starting of the IDDP (the Iceland Deep Drilling Project) which discovered that the hot magma was only 2.1 kms below ground, which is very close to the earth's surface.

My current opinion

Although the Krafla and associated volcanic fissure systems are extremely large and definitely still active, I feel that this sort of volcano has regular eruptions and then some lava fountains erupt followed by lava flows, it seems that these eruptions are more sedate (if such a word can be used for a volcano) compared to the explosive destructive forces at play with Hekla, Laki and Bardarbunga and other more effusive volcanoes. Still having said that, historically Krafla has erupted 29 times in recorded historical times.

Personally (although it must be remembered that I am not a qualified volcanologist and these are only my own personal thoughts on the matter having perused the data available) I do not think that Krafla will be the one to cause chaos over Europe, having said that once again I hope I am not proved wrong.

Danger Level: 4 (out of 10)

Eruptability Level: 35 % probability of an eruption in the next 10 years.

18. Laki or Lakagigar (Warning Level 1 risk)

By kind permission of Mr Oddur Sigurdsson jardfraedingur (geologist) Icelandic Meteorological Office, Reykjavik, Iceland

"An aerial view of the Laki volcano in the Grimsvotn volcanic fissure system in Central Iceland. When this fissure system erupted in 1783 Europe was covered in a poisonous yellow haze called "the Laki haze" and it is estimated that about 1 million people died in Europe between 1783 and 1784. This is definitely one of the most, if not the most dangerous volcano in the whole of Iceland and if it erupts again we had all better be well prepared for the dire consequences which are bound to follow."

Style: Fissure volcanic system
Location: S E Iceland near Grimsvotn and South of Bardarbunga. Not far from the valley of Eldgja and the small village of Kirkjubeaejarklauster.
Height: 1,725 m (5,659 ft)
Last Erupted: 1783 – 1784

Laki or Lakagigar (or also called Loki perhaps named after the Norse god of mischief on some maps) is one of the baddest boys on the block. As we previously discussed the Laki eruptions of 1783 – 1784 affected most of Europe as far as Prague and possibly killed 1 million people and caused the French Revolution, not bad for small fissure in the side of a mountain in S E Iceland, definitely top of the Premier League of dangerous volcanoes. This is one hot potato that you don't want to hear about one morning over your cornflakes, that Laki has erupted and a yellow toxic cloud is 72 hours away from Manchester, that would definitely depress me even more than the normal Monday morning depression.

In 1783 as we discussed earlier, technically the mountain did not erupt, several fissures opened in its sides and actually Laki is definitely part of the Grimsvotn system which also includes the Thordarhyrna volcano. It sits on the mid Atlantic spreading ridge and the fissures under Laki run S W to N E like many other volcanic features in this area.

The system erupted over an 8 month period in 1783 – 1784 from the Laki fissure and from the adjoining Grimsvotn volcano pouring out an estimated 14 km3 (3.4 cu miles) of basalt lava and clouds of poisonous hydrofluoric acid /

sulphur – dioxide compounds that killed over 50 % of Iceland's livestock population which led to famine which killed about 25 % of Iceland's population.

The Laki eruption and its aftermath have been estimated to have killed over 6 million people globally making it the most deadly natural disaster in recorded historic times.

1783 – 1784 Eruption

On 8th June 1783, a fissure with 130 craters opened with phreatomagmatic explosions due to the subsurface groundwater interacting with the rising flood basalt magma. Over a few days the eruptions became less explosive, strombolian and later more akin to Hawaiian in their character, with high rates of flowing lavas. This volcanic event is rated as VEI 6 on the Volcanic Explosivity Index, however the 8 month ejecting of sulphuric aerosols into the atmosphere resulted in one of the most devastating effects on both human habitation and climactic events within human recorded history.

This eruption which was also known as the Skaftareldar (Skafta Fires) or Sioueldur, produced an estimated 14 km3 (3.4 cu miles) of basalt lava and the total volume of tephra ejected was 0.91 km3 (0.2 cu miles). Lava fountains reached heights of 800 – 1400 m (2,600 – 4,600 ft). In Britain the summer of 1783 was known as the "sand – summer" due to the amount of sand like ash that continued to fall over The British Isles that year. The dangerous gases were carried by the convective ash cloud column up to altitudes of 15 kms (10 miles) high into the atmosphere.

Although the main eruption continued until 7th Feb 1784 most of the actual lava erupted within the first 5 months of the main eruption. Grimsvotn, which is connected to the Laki fissure was also erupting from 1783 – 1785. The ejection of various gases includes an estimated 8 million tons of hydrogen fluoride and a further estimated 120 million tons of sulphur dioxide, this gave rise to what was known then as the "Laki Haze" which spread across Europe causing havoc and mayhem.

Repercussions for Iceland

The resulting difficulties faced by the Icelandic population at that time was known as "The Mist Hardships," 20 – 25 % of the population died of famine and fluorine poisoning after the eruptions had ceased. 80 % of sheep, 50 % of cattle and 50 % of horses died due to dental and skeletal flourosis from the 8 million tons of hydrogen fluoride that were released. It is sad to relate, however most of the coastal settlements which were fishing communities were still exporting dried and salted fish to Europe and North America whilst their brethren were dying of starvation in the Central Highlands and elsewhere in Iceland at that time as there was no central authority to co ordinate any rescue or relief works.

Consequences for Europe

It has been calculated that more than 120 million tons of sulphur dioxide was erupted into the atmosphere, about 3 times the total annual European industrial output for 2006, and equivalent to a 1991 Mount

Pinatubo eruption every 3 days. This mass of sulphur dioxide during abnormal weather patterns caused a thick haze or fog to spread slowly across Western Europe resulting (as we stated earlier) in thousands of deaths across Europe that summer. Sadly for all concerned the summer of 1783 was one of the hottest on record and a rare high pressure zone settled over Iceland caused the main winds to blow to the South East. The poison gas clouds drifted firstly to Bergen in Norway, then across Europe to Prague in the Kingdom of Bohemia (now part of the Czech Republic) by 17th June, Berlin 18th June, Paris by 20th June, Le Havre by 22nd June and to Great Britain by 23rd June. The fog was so thick in places that ships were unable to navigate and had to stay in port and the sun was described as "blood covered".

If one inhales sulphur dioxide gas it causes the victim to choke as their internal soft tissues swell up. Parish records across Britain and France show a massive increase in deaths from respiratory causes from that summer.

The temperatures across the continent became increasingly unbearably hot causing massive thunderstorms to break out with large hailstones mixed into them which are reputed to have killed cattle in several locations. The haze finally dissipated in the late autumn of 1783.

However next came the freezing winter of 1784, a certain Gilbert White at Selborne, Hampshire reported that there had been 28 straight days of frost lying on the ground. This extreme cold is estimated to have contributed to a further 8,000 deaths in the UK alone and we do not know how many perished on the continent. There were reports of severe flooding in the spring of 1784 from Germany and other parts of Europe.

Over the next few years across Europe extreme differences in weather patterns were noted, there was a surplus harvest in France in 1785 that caused much poverty as grain prices slumped for the rural population, then droughts and poor harvests followed pushing many peasants into the French city's and as described earlier this was one of the main contributory factors for the French Revolution of 1789. Apart from Laki, Grimsvotn was also erupting from 1783 to 1785 and combined with a strong El Nino effect from 1789 – 1793 weather patterns were altered considerably by these eruptions.

The effects upon North America

1784 was one of the most bitterly cold winters ever to hit the North American Continent in recent historical times. It encompassed one of the continually longest periods of sub – zero temperatures in the New England area and the largest amount of snow ever seen in New Jersey. The Chesapeake Bay froze over for many months and there was ice skating in Charleston Harbour, a huge snow storm hit the deep south of the fledgling USA and the Mississippi River froze over at New Orleans and ice was seen in the Sub Tropical area of the Gulf of Mexico.

Effects Worldwide

It is estimated that the cooling effect from the aerosol injection of reflective sulphur dioxide in the upper atmosphere in 1783 – 1784 affected rainfall patterns across the globe and there were terrible droughts in India, Egypt and as far away as Japan, it is thought that up to 6 million souls perished worldwide due to this catastrophe.

Gilbert White recorded his perceptions of the event at Selborne, Hampshire, England.

"The summer of the year 1783 was an amazing and portentous one, and full of horrible phenomena; for besides the alarming meteors and tremendous thunderstorms that affrighted and distressed the different counties of this kingdom, the peculiar haze, or smokey fog, that prevailed for many weeks in this island, and in every part of Europe, and even beyond its limits, was a most extraordinary appearance, unlike anything known within the memory of man. By my journal I find that I had noticed this strange occurrence from June 23 to July 20 inclusive, during which period the wind varied to every quarter without making any alteration in the air. The sun, at noon, looked as blank as a clouded moon, and shed a rust- coloured ferruginous light on the ground, and floors of rooms; but was particularly lurid and blood-coloured at rising and setting. All the time the heat was so intense that butchers' meat could hardly be eaten on the day after it was killed; and the flies swarmed so in the lanes and hedges that they rendered the horses half frantic, and riding irksome. The country people began to look, with a superstitious awe, at the red, louring aspect of the sun."

Benjamin Franklin recorded his observations in a 1784 lecture:

"During several of the summer months of the year 1783, when the effect of the sun's rays to heat the earth in these northern regions should have been greater, there existed a constant fog over all Europe, and a great part of North America. This fog was of a permanent nature; it was dry, and the rays of the sun seemed to have little effect towards dissipating it, as they easily do a moist fog, arising from water. They were indeed rendered so faint in passing through it, that when collected in the focus of a burning glass they would scarce kindle brown paper. Of course, their summer effect in heating the Earth was exceedingly diminished. Hence the surface was early frozen. Hence the first snows remained on it unmelted, and received continual additions. Hence the air was more chilled, and the winds more severely cold. Hence perhaps the winter of 1783-4 was more severe than any that had happened for many years.

The cause of this universal fog is not yet ascertained [...] or whether it was the vast quantity of smoke, long continuing, to issue during the summer from Hekla in Iceland, and that other volcano which arose out of the sea near that island, which smoke might be spread by various winds, over the northern part of the world, is yet uncertain." (According to contemporary records, Hekla did not erupt in 1783; its previous eruption was in 1766. The Laki fissure eruption was 45 miles (72 km) to the east and the Grimsvotn volcano was erupting circa 75 miles (121 km) north east. Additionally Katla, only 31 miles (50 km) south east, was still renowned after its spectacular eruption 28 years earlier in 1755.)

The Reverend Sir John Cullum of Bury St Edmunds, Suffolk, Great Britain recorded on the 23rd June 1783, the same date that Gilbert White noted the beginning of the unusual atmospheric phenomena, in a letter to Sir Joseph Banks, then President of The Royal Society.

"...about six o'clock, that morning, I observed the air very much condensed in my chamber-window; and, upon getting up, was informed by a tenant that finding himself cold in bed, about three o'clock in the morning, he looked out at his window, and to his great surprise saw the ground covered with a white frost: and I was assured that two men at Barton, about three miles (5 km) off, saw in some shallow tubs, ice of the thickness of a crown-piece."

Sir John goes on to describe the effect of this 'frost' on trees and crops:

"The aristae of the barley, which was coming into ear, became brown and withered at their extremities, as did the leaves of the oats; the rye had the appearance of being mildewed; so that the farmers were alarmed for those crops...The larch, Weymouth pine, and hardy Scotch fir, had the tips of their leaves withered."

Sir John's vegetable garden did not escape; he noted that the plants looked *"exactly as if a fire had been lighted near them, that had shrivelled and discoloured their leaves."*

My current opinion

This volcano speaks for itself, there were other spectacular eruptions in the 18th Century from Laki in 1766 and possibly other eruptions which may not have been recorded. I would have to state that after reading all the various reports that in my opinion a Laki eruption which was similar to the 1783 – 1784 eruption would pose the biggest threat of all to all the inhabitants of Europe and North America (depending on which way the prevailing winds were blowing). If you heard on CNN or Sky News that Laki was in full eruption then everyone had better heed my instructions and lock yourself in your home for about a month to avoid the worst case scenarios which will play out and do not have to be repeated here again.

My fervent hope is that the Governments of Europe heed my and the President of Iceland's warnings and immediately start laying in all manner of provisions and water to feed and water the population for up to six months in the worst case situation, doing nothing about these threats is not an option. It is not as if these volcanoes erupted in Roman times or prehistoric times, Laki erupted just 230 years ago, just after the founding of the USA and during the early part of the Industrial revolution in the UK, it could and it will happen again and unfortunately for everyone in its path, due to the chemical makeup of Laki's lava's and gases it does eject poison gas which will have a terrible effect on anyone caught out in these killer fogs.

I doubt whether hundreds of thousands of farm labourers or city dwellers will drop dead in the streets as people will heed my and the Governments warnings to stay indoors until the fog has dissipated, however unwary travelers, people who live in the streets and many others could face life threatening injuries and possibly death unless they get shelter and food fairly quickly after the initial eruption.

Danger Level: 8 – 9 (out of 10)

Eruptability Level: 55 % probability of an eruption in the next 10 years.

19. A brief list of Smaller and less dangerous Icelandic Volcanoes			
Name	**Type**	**Last Erupted**	**Danger 1 – 10**
i. Langjokull	Fissure system	1000 ACE	2
ii. Ljosufjoll	Fissure system	1148 ACE	2
iii. Oraefajokull	Strato volcano	1728 ACE	4
iv. Prestahnukur	Rhyolitic Dome	7500 BCE	3
v. Reykjanes	Sub surface lava fields	Holocene Era	3
vi. Reykjaneshryggur	Not Known	1970	2
vii. Snaefellsjokull	Strato Volcano	200 ACE	2

viii. Surtsey	submarine volcanic system	1963	2
ix. Tindfjallajokull	Strato Volcano	Holocene Era	2
x. Tjornes	Not Known	1868	2
xi. Torfajokull	Strato Volcano	1477 ACE	3
xii. Tungnafellsjokull	Strato Volcano	Holocene Era	2
xiii. Vatnafjoll	Fissure System	800 ACE	2
xiv. Peistareykjarbunga	Not Known	750 BCE	2
xv. Porolfsfell	Tuya volcano	Not Known	?

CHAPTER 8

Conclusions

We have now discussed at some length what the various scenarios could be if and when another Icelandic Volcano erupts in the very near future. We have looked at the risks of other natural disasters compared to volcanic eruptions, we have discussed what governments and individuals should do to minimize the risks, we have looked at the sort of food, heating, clothing, medicines people should start putting away for the day that might come soon.

We have also looked at an example of tourists who are trapped abroad and shown how difficult it might be for them to return home. We have discussed food distribution by governments, travel disruption and many other relevant issues. In the previous chapter we have tried to highlight which volcano might be the one to cause all this chaos and as you can now see a few of the bigger active ones seem to be the ones to watch closely, namely, Hekla, Katla, EJ, Bardarbunga, Grimsvotn and Laki amongst quite a few others. The good news is that various Icelandic and International Institutes of Volcanology are now keeping a watchful eye over many of these volcanoes and although I doubt if we will get many weeks of warning, we should get a few days to prepare for the worst, before any major eruptions occur. Having said all that there have been instances in the past when some of these volcanoes erupted with little or no warning. I suppose that the UK and Europe is about 1000 miles away so in any event we would get a few days warning when one of them blows its top.

What should we all do now? Well, I feel that we should be lobbying our MP's and Ministers and Prime Minister to start to take some urgent action in preparation for the worst case scenario. I have already written to David Cameron and Boris Johnson (Mayor of London) and have not had the courtesy of a reply yet from the PM's office, however I did receive an e-mailed response from The Mayor of London's Office advising me that the Mayor's Office was deeply involved in the COBRA project which Is basically a hierarchical list of emergency committees that will meet with members of the police, fire brigade and ambulance services and maybe the army to co ordinate any response from any external emergency and threat to the UK and London.

For example when the 7 / 7 bombers struck London it would have possibly been COBRA that coordinated the authorities responses to those attacks. On paper it sounds like a sensible thing to do, to have a coordinated response to any threat and I am sure that the French, German, Spanish and Dutch Governments have a similar organization. However it is all well and good having all these committee's organizing responses, however in the event of a major volcanic eruption what will be needed will be food supplies, possibly breathing apparatus (for a Laki style gas eruption), community centre's, food distribution points, protected water supplies and a strategic oil and fuel reserve as well as plans to keep the power on and water and sewage plants running at the same time. I doubt if all of the above points have been rationally thought about and an actual action plan made. If there is such a plan why has it not been made public? We are not talking about a terrorist threat, this will be the result of a natural disaster and we need to know that our Government is prepared.

I have been informed that the Army does have stockpiles of emergency rations and as described elsewhere in this book, there are historical stockpiles of canned foods dating back 30 – 50 years, however I doubt that these meagre supplies will be able to feed 62 million people for about six months (at the worst case scenario). The Government should be building up some strategic reserves of long life foods, perhaps just basic foods, such as meats, beans, sweet corn, dried powdered milk, safe water supplies, pasta, grain (for bread and feeding to livestock) and similar basic foodstuffs, together with campingaz canisters and burners and heaters and stores of paraffin and paraffin heaters in case the power grids are blacked out for some time. Stocks of camp beds, blankets, sheets, pillows and water treatment pills, medicines and some of the basic materials that I have mentioned earlier in my book would also be useful.

It might possibly be sensible for the Government to get in touch with the American Website (Survival Acres) who have foods which are reported to last for up to 40 years, the Government could buy a few ship loads and store them away somewhere for an emergency and that could be a solution to at least some of the problems.

Recently in the news, much discussion has taken place regarding cuts to many services in the UK and Europe following the International Credit Crunch, many hundreds of thousands of civil servants and local authority employees might be losing their jobs and many strikes have been threatened, currently in France there is a strike of oil terminal workers amongst others and this has affected petrol stations across France with motorists unable to buy fuel in some areas, imagine now on top of this chaos a volcanic eruption in Iceland, the domino effect would be catastrophic as services already weakened by strikes and cuts to Government employees might make the Governments response to this latest threat much slower and weaker due to lack of resources from strike bound materials and supplies, especially fuel and oil supplies.

I am not advocating banning strikes, however authorities need to be aware of the multitude of dangers that could affect their response to a serious emergency should it arise.

Sadly the current difficult economic times faced by most inhabitants of Europe means that people are more worried (quite rightly) about their jobs, family and other more basic needs of keeping a roof over their heads. Many might feel that the level of risk from an erupting Icelandic Volcano is small and even if it were to happen that it wouldn't affect their way of life and therefore it was of minimal importance. I would beg to differ, in my opinion apart from 2 World Wars in the 20th Century and the Spanish Influenza Pandemic of 1918 the next major threat to life in Europe, in my opinion is a massive

Icelandic Volcanic Eruption. In 1783 – 1784, 6 million people died worldwide due to Laki's activities. This was only 230 years ago, however before the EJ eruption of March – April 2010 I had never even heard of Laki or these other large and dangerous volcanoes fitfully slumbering over in Iceland.

I do find it strange that some major events in history seem not to be remembered or recorded whilst other such as the Jacobite Rebellions of 1715 and 1745 and the Battle of Waterloo in 1815, the sinking of the Titanic in 1912 seem to be indelibly stamped on the nations psyche. In this instance I would surmise that the average Englishman living in 1783 – 1784 would have seen this poisonous yellow fog as some sort of natural anomaly, like the killer smog's of the 1940,s and 1950,s before they brought in the clean air acts. Therefore the average citizen in the Eighteenth Century wouldn't have known about the eruption of Laki 1000 miles away in Iceland and at that time no explanation was given as to how the fog came to envelop the UK, so it wasn't seen as the global killer that we now see it to be.

Although this book is primarily targeted towards Icelandic Volcanoes, we should not forget that approximately three million Italians live within striking range of Vesuvius. which recently has started rumbling again, then there is Etna and Stromboli not far away, these European volcanoes would also have a devastating effect on Italy and other parts of Europe should they have a serious eruption and currently I understand that the Italian Government is trying to encourage people actually living on Vesuvius' flanks to move to safer areas, however there are potentially three million souls currently at risk from a major Vesuvian eruption.

My attempts in exposing the current levels of risks to European citizens are not meant to be alarmist in any way, I personally don't think the Icelandic Volcanoes unlike the Yellowstone super volcano are going to change our civilization in the long term, I do feel though that if one of the big boys goes off then there will be a tremendous disruption caused to everyone's daily lives, jobs will be at risk as communications, distribution services and service industries will all be adversely affected by the ongoing chaos and mayhem caused by the eruptions. One of the biggest casualties economically will probably be the airlines and travel companies, as if there are no airlines flying over the Atlantic and Europe for six months many national carriers and independent airlines might fold.

If a huge ash cloud was engulfing Europe the Transatlantic Carriers may have to fly from Miami or Atlanta to Casablanca or Tunis and these airports that may not have the latest computer aided navigational aids would probably soon by swamped by the number of planes trying to land there. Passengers arriving in Marrakesh or Casablanca from say New York, whom wanted to get to Berlin, would then have to wait for ferries, trains and coaches to complete their journeys if at all.

If there was a poison fog alert over Europe the travelers would be stuck in Casablanca for many weeks maybe before the authorities allowed any onward passage?

Passenger chaos would be one issue, another major issue would be logistics.

If Sainsbury's Supermarkets ran out of foods, let us discuss just one foodstuff, say tinned sweet corn for arguments sake, their computerized ordering system would then send an e-mail or fax to the supplier, say in South Carolina USA for another fifty pallets of tinned sweet corn to be shipped to the UK urgently. The producer would (if he had the capacity) then organize four HGV lorries to take perhaps four containers to the container port of Delaware. However the poor overworked shipping agent would have also received countless other orders from supermarket chains in Europe for urgent orders of foods and other commodities and there would suddenly emerge a huge bottleneck where there would not be enough space on the container ships currently in service to handle all these urgent extra requests for shipping. Currently due to the world's economic slowdown many container ships and tankers have been mothballed and put into storage until the world's economy picks up. One simply cannot row a crew out to Falmouth Harbour, turn an ignition key and sail a 60,000 ton container ship over to Delaware to pick up containers, to get a ship out of mothballs and check the engine and hull would take perhaps a month, so all these increased orders to re supply stricken supermarkets would simply back up in the shipping agents offices.

It is true that Sainsbury's, Tesco's and Asda would hunt around the globe for alternative suppliers, however from Taiwan to Cape Town everyone would have the same problem, inventories have been cut down due to the recession, so producers will then have to rush out and get the raw material, in the case of sweet corn, this will be maize, however maize has a growing season, you cannot just conjure up 50,000 tons of maize, strip the sweet corn off, cook it, can it and ship it off to Blighty in a few days. The lead in period for all these new orders in a normal market is probably one to two months ahead. Normally Sainsbury's shelves will empty and an order goes to head office, there are some stocks in depots in the UK, however the computers try to forecast usage trends and will only re order from suppliers when central depots start to run down their inventory (stock of foods in this case).

In the sort of crisis I am contemplating, the supermarkets will sell out immediately, then they will attempt to restock from their UK depots, those foods will again fly off the shelves as everyone panic buys, then the long delays and bottlenecks will kick in, the length of time taken, even if drivers for lorries and docking space is available could be several months potentially, unless the governments involved commandeer ships and friendly governments like the Americans pull a few strings to ensure any stocks of emergency foods available in the US can be diverted to the UK. Whilst all these actions are being taken the general chaos of key personnel being stuck in the wrong place, possible communications and power break downs and many other insurmountable problems, will make re supplying supermarkets a Herculean task. However for those sensible citizens who bought this book, logically thought about the consequences and made some provision for what might occur should the worst happen, all of the above although worrying will not turn into a life and death struggle for survival. As previously discussed, I personally do think governments will co operate with each other and some basic distribution of necessities, like bread, potatoes, some canned foods and water might occur after a couple of weeks, however why risk the unknown, if 70 % of the UK population takes my advice and stores enough food, water and other supplies for just 3 months then this will also ease the pressure on the Government and supermarkets to re supply the population immediately.

One of the other consequences of people not being able to travel freely and people not going to work regularly will be the fact that many people will start to run out of money fairly soon and will again be reliant on Government hand outs. Many businesses are currently facing financial pressures due to Government cutbacks and the credit crunch, if no one is shopping in the high street and no one is going to work then what happens to trade and commerce generally, it all seizes up and everything has a knock on effect. If firms owe money to other firms and cannot pay, then further pressure is put on an already creaking system. The Government will probably have to announce some sort of moratorium on mortgage payments and commercial and personal bank loan payments, as no one will be able to keep paying thousands of staff, who firstly are not at work and secondly, that their income stream has dried up as everyone is concentrating on coping with the emergency. It may be that the Government has to re introduce some form of rationing and state subsidized food, clothing and heating support systems for some months even after the emergency is over.

Similarities to how we coped during WWII do keep on surfacing here, the British population had ration books, the Government had opened subsidized canteens like NAAFI canteens, people were advised to dig up their own gardens and parks to plant allotments to grown their own vegetables and fruit. People kept chickens in their gardens. The authorities made synthetic copies of scarce foods, i.e. powdered eggs, powdered custards, Ersatz Coffee (The Germans used this more than the Brits). Strangely enough the wartime diet was actually probably healthier than most people's diets now, there were very few sweets around, no cane sugar, we had to make beet sugar and use honey as a sweetener, and kid's teeth were probably healthier too, no chocolates or sweets for the kids until the Yanks arrived over here in late 1942 with all those dam Hershey Bars and Bazooka Joe chewing gum. Our diet was basically a small amount of beef or chicken, unless you were an RAF pilot who received the full English Breakfast before scrambling into his Spitfire and racing up to do battle with the Luftwaffe over Southern England, the rest of us had to make do with bread, potatoes, leeks, parsnips, turnips, carrots, swedes and many other root vegetables, the occasional apple, occasional pasta and no fruits like banana's, pineapples, mango's for years during the war. Dieticians looking back on this period think that the British public were a lot healthier than they are nowadays.

Clothes were recycled in shops where you brought your own clothes into the shop, handed them in and were then allowed to buy or loan other peoples clothes (hopefully washed and / or dry cleaned firstly). Petrol was rationed for war and emergency use only and a red dye was put in the petrol so that the police could check and see if vehicle owners were using black market fuel or not. The public had to rely on public transport, the railways were nationalized and people used buses, trams, the underground and trains to get around, this will probably have to happen again.

However what was one of the most important features of the war years and probably not seen before or since was a sense of camaraderie and the thought that everyone was in this together and together as a nation we pulled through the darkest hours of the last war. I am not sure that David Cameron can be compared to Winston Churchill, however funnily enough we did have a coalition government throughout the war and they did have to preside over some pretty drastic cuts and changes to what the British Public were used to, so quite a few similarities will come to the fore. I am hoping that the British tradition of fair play and carrying on as normal will come to be standard practice in the next trial of the British people and our European Cousins across the Channel. If we all stand together our British ingenuity and resilience will see us through any crisis that looms ahead.

When will it happen?

This is the $ 65,000 question? There have been 41 eruptions from 1902 to 2010 when EJ blew its top, there is no way of knowing for certain which volcano and at what time will it erupt. All we can do is monitor the Volcanology Institute of Iceland and the Nordic Vulcanology Institute's website and post any news on our website www.icelandicvolcanoes.co.uk. Looking at the list, most of our old friends seem to be on it;

LIST OF ERUPTIONS OF ICELANDIC VOLCANOES IN THE 20th & 21st CENTURIES

Askja: 1921, 1922 (×2), 1923, 1924, 1927, 1929, 1961,

Austan Hekla: 1913

Eldey: (Northeast of Iceland): 1926.

Eyjafjallajokull: 2010.

Grimsvotn: 1902, 1922, 1933, 1934, 1938, 1983, 1998, 2004.

Gjalp: 1996

Hekla: 1947, 1948, 1970, 1980, 1981, 1991, 2000.

Heimaey: 1973.

Katla: 1918.

Krafla: 1975, 1977 (×2), 1980 (×3), 1981 (×2), 1984.

Landeyjar (sub aquatic eruption 5kms south of Iceland) : 1973.

Pordarhyrna: 1903, 1910.

Surtsey: (partly sub aquatic) 1963 – 1967.

Worryingly Laki is not on it and if there is one Icelandic Volcano to really keep a watch on it is the Laki area of the island due to the history of poison gas eruptions in the past. The longer it doesn't erupt the more powerful the eruption will be when it finally does go off, that is the theory at least. What I can say with some certainty is that there is an extremely strong possibility of one or more of the above volcanoes erupting over the next five years. Strangely apart from the 1950's, when they were having a bit of a breather, Iceland's volcanoes have erupted in every decade since the turn of the 20th Century, probably apart from Hawaii, Iceland is one of the earth's most active volcanic zones, sitting atop a hot plume of magma rising from deep within the earth's mantle and being on a hot spot and sitting abreast the Mid Atlantic Spreading Ridge. In fact on the plus side Iceland is almost totally constructed from volcanic rocks over the ages, on the flip side I wouldn't care to be a sheep or a cow on Iceland, chances are before you head for the slaughter house you will have died from fluorine poisoning, sulphur dioxide poisoning or been burnt alive from the heat of an eruption. If one were to be any animal on Iceland, I would rather be a seal, as Iceland used to have the most plentiful stocks of fish across the whole North Atlantic.

Finally, whilst I am not advocating that everyone puts on a billboard stating that "the end of the world is nigh" and walk up and down Oxford Street, although I have seen quite a few people doing that lately, and I do not subscribe to 2012 being the end of the world, I would say that if people start building up their emergency reserves of food, water and other supplies as I have listed earlier, bit by bit, maybe a crate of tinned food purchased extra every fortnight then when some really bad news does come across the BBC or CNN News Flash one night, you will know that you will have one less thing to worry about than most of the population that didn't read this book. I will probably have to go out and purchase a lot of extra canned foods and bottled water as everyone now knows that I wrote the book so I will have plentiful supplies and furthermore my wife is from a large family of nine sisters and brothers and I can guarantee that they will be tap tapping on my door once the balloon goes up, and however many times I could have said "I told you so" their little outstretched hands and that look in their large eyes will tell me that my wife has already agreed to give all of our meagre supplies away. However don't worry about me because I have read my book and noted the comments about having a secret cache of food and water, and unless I am having my nails removed under torture, (which I suppose is a possibility) I will not be revealing the secret of the secret cache's address (just to annoy my sister in laws, there might be several secret cache's, but please nobody tell them).

I will also definitely be travelling to Iceland to actually go and see as many of these volcanoes as I can in 2011 and will be taking many digital photos of the volcanoes discussed in this book and will be adding the photos to the official website: www.icelandicvolcanoes.co.uk so that readers of this book can see the volcanoes mentioned in the book as they are today. Let's hope the balloon does not go up whilst I am in Iceland as if I am trapped there what will happen to my wife's brothers and sisters and how will I get access to my secret cache of supplies? I will be reduced to living on herring and poisoned sheep and cow for some months which is not an exciting prospect although I have been reliably informed that the Icelandic race are extremely friendly and hospitable which is rather strange considering that they have all been living on this large gunpowder keg of an island which seems to be in a continual state of eruption more or less every couple of years.

I am hoping to do some more research into the history and background of the Icelandic volcanoes when I am in Iceland and will try and write some addendums, and revisions to a possible 2nd Edition of this book with various updates, some of the revisions I will try and précis and put on the website as it would be grossly unfair to my faithful readers to have to go out and buy this book again just to read a ten page addendum and or revision.

As a final note to this book, I am not related to anyone at Tesco's, Sainsbury's, Asda, Lidl, Morrison's, Waitrose or any other supermarket chain. I do own 40 shares in Tesco though, I would advise those of my readers who are into finance and share dealing that it might be a good bet to buy some supermarket share's once this book is published, however after a few months perhaps sell those share's before any eruption wipe's the smile off the supermarket bosses face for a while. Seriously though please remember shares can go down as well as up and the best place, in my opinion, for any spare money is under your mattress, especially if power cuts mean that cash machine's won't be dispensing monies and bank tellers might not be able to get to their offices.

Currently at the time of completing this book, mid November 2010 a volcano called Mount Merapi has just erupted in Indonesia on the same day that a massive tsunami from a subterranean earthquake struck the country, in a very similar area to the massive 2004 Boxing Day tsunami, there are also some very active strato volcanoes erupting in Kamchatka, Far Eastern Russia and several other volcanoes and earthquakes active around the world currently. I suppose because I am interested in these phenomena that I do take a mental note of what is happening tectonically speaking around the world and I have noticed through news reports and the media generally that there does appear to be an increase in seismic and volcanic activity currently, it may be that due to the convection currents in the mantle under the earth's surface that there are cyclical periods when this sort of activity increases and decreases, perhaps even linked to sunspot activity on the sun's surface, no one really knows what starts off a cycle of crustal movement and tectonic activity, however if I was a gambling man, which I am not, I would put a fair sized wager that before the year 2016 I would expect to see a fairly major eruption in Iceland or Europe or a large eruption elsewhere in the world that directly effects Europe's weather and climate and economy occurring.

The speeding up of the melting of the Greenland Ice Cap and to a lesser degree of the Icelandic Glaciers and icecaps, as I mentioned before, the weight of all this ice, especially I think the Greenland Ice Cap could act like taking a large stone off of a fountain, the magma which has been kept bottled up under these ice sheets could start causing crustal displacement as the magma wells up from the mantle and the plume under Iceland. In some parts of Iceland magma is only a couple of kilometers under the surface. As global warming increases so the pressure on the champagne cork theory also increases, if you add to this scenario a situation where The North Atlantic Conveyor System, which to you and me means the Gulf Stream, which is a massive current of warm water which comes all the way from the Equator via the Caribbean up through the Atlantic and ends up off Northern Norway, this system could start to fail if too much fresh water comes off the Greenland Ice Sheet so causing the sea waters salinity to go down, this in turn stops the cooling water to sink and go back down to the equator to start heating up again. If this system slows and stops, then this alone could tip the Northern and Western parts of Europe the UK and Ireland into a mini ice age and if you add to that a few Icelandic volcanoes popping their champagne corks so to speak, you have a recipe for a very chilling and dangerous scenario for all of us who live in Europe and possibly North East USA as well. There are of course too many differing scenarios for me to mention all of them here, suffice to say that any of them would be bad enough and a combination of them would be positively lethal.

You have been warned, please take heed of this book and take the appropriate actions discussed in it and please check out my website for further information and advice regarding all the matters that have been discussed in this book. www.icelandicvolcanoes.co.uk. Let us all hope for all of our sakes and our children's sakes that I am wrong and I lose my bet and that nothing untoward happens to us here in Europe or the rest of the world for a long time to come.